PAINLESS
Fractions

Alyece B. Cummings
illustrated by Laurie Hamilton

BARRON'S

All inquiries should be addressed to:
Barron's Educational Series, Inc.
250 Wireless Boulevard
Hauppauge, New York 11788
http://www.barronseduc.com

International Standard Book No. 0-7641-0445-4
Library of Congress Catalog Card No. 98-23068

Library of Congress Cataloging-in-Publication Data
Cummings, Alyece B.
    Painless fractions / Alyece B. Cummings : illustrated by Laurie Hamilton.
        p.   cm.
    Includes index.
    Summary: Defines and demonstrates fractions and the terms associated with them, explaining how to add, subtract, multiply, and divide them and providing advice for avoiding pitfalls in working with them.
    ISBN 0-7641-0445-4
    1. Fractions—Study and teaching (Elementary)—Juvenile literature.   [1. Fractions.]   I. Hamilton, Laurie, ill.
    II. Title.
QA117.C86   1998
513.2´6—dc21   98-23068
                                                              CIP
                                                              AC

PRINTED IN THE UNITED STATES OF AMERICA
9 8 7 6 5 4 3 2 1

This book is for anyone who needs a refresher on fractions and wants to have fun while reviewing.

Acknowledgments

A world of love and thanks go out to my husband, Steve, and my daughter, Ilene. Their tireless typing efforts, computer expertise, and loads of patience helped to make this book become a reality.

Thanks also to my editors, Amy Van Allen, Heather Miller, and Marcy Rosenbaum, my illustrator Laurie Hamilton, the computer department that deciphered my diskettes, and the rest of the wonderful staff at Barron's. In addition, a special thanks goes to Grace Freedson for having faith in me.

TANTALIZING TABLE OF CONTENTS

BRIGHT BEGINNINGS

This book is a fun refresher book on fractions. Some of my students once mentioned that they would like a math book that was on an easy level to understand, wasn't boring, and explained things step-by-step, so I came up with this.

Chapter One is an introduction into the world of fractions. We'll define a few terms, change from mixed numbers to improper fractions and vice-versa, reduce and simplify fractions, and build equivalent fractions.

Chapter Two compares fractions to fractions, fractions to decimals, and fractions to percents. Which has the greatest value?

Chapter Three moves into multiplication with proper fractions and mixed numbers.

Chapter Four dives into division with its rules for proper fractions and mixed numbers.

Chapter Five adjusts to addition with and without those common denominators, proper and improper fractions, and more.

Chapter Six submerges under subtraction. Besides the common denominators, like we used in addition, we also need to learn to "borrow."

Chapter Seven wades into word problems. Math students are seldom overheard saying, "Gee, I really love word problems. They're my favorite part of math." Hopefully here, with some pointers and practice, they'll be just as doable as non-word problems.

Chapter Eight rolls over ratios, rates, and proportions. Here we'll compare ounces to ounces, figure out miles per hour, and solve for missing parts.

Chapter Nine exercises with exponents. We'll practice problems with those little numbers and see how they change the value of a number.

Chapter Ten puts it all together with the order of operations. "What do I do first in the problem?" After practicing problems in Chapter Ten, all your questions will be answered.

Chapter Eleven finishes with practice, practice, practice, and even some puzzles to play around with. Review, refresh, and enjoy!

CHAPTER ONE

Intriguing Introduction

What did one fraction say to the other fraction?

"You've seen the better part of me."

What is a fraction?

A fraction is a part of a whole.

1 whole pizza

1 whole pizza can be divided into different fractional parts.
For example,

Important terms

Here are some terms that will only tax your brain a fraction:

Fraction Parts

$\dfrac{1}{2}$ ← Numerator
← Denominator

Is a proper fraction one that does everything right?

No, a **proper fraction** is one where the numerator is smaller than the denominator, like $\frac{3}{4}$ or $\frac{1}{2}$.

Is an improper fraction one that does not have manners?

No, an **improper fraction** is one where the numerator is larger than or equal to the denominator, like $\frac{4}{3}$ or $\frac{6}{5}$ or $\frac{8}{2}$ or $\frac{3}{3}$.

Is a mixed number one that is stirred a lot?

No, a **mixed number** contains a whole number part and a fraction part, like $3\frac{2}{3}$ or $4\frac{7}{8}$.

MIXED NUMBERS AND IMPROPER FRACTIONS

How do you change mixed numbers into improper fractions and vice versa?

Is it a magic trick? Not even a famous magician can change numbers into other forms without following certain rules.

(1) To convert from a mixed number to an improper fraction:

Multiply your denominator by your whole number.
Add the result to the numerator.
This number becomes the numerator and the denominator remains unchanged.

CORRECT:

$3\frac{4}{5}$ $5 \times 3 = 15$

$15 + 4 = 19$

The improper fraction is $\frac{19}{5}$.

INCORRECT:

$3\frac{4}{5}$ $5 \times 3 \times 4 = \frac{60}{5}$

or

$4 \times 3 + 5 = \frac{17}{5}$

(2) To convert from an improper fraction to a mixed number:

Divide the denominator into the numerator. This answer becomes your whole number. The remainder becomes the numerator of your fraction, and the original denominator stays as the denominator of your fraction.

RIGHT:

$\frac{14}{3}$; 14 divided by 3 equals 4, with a remainder of 2.

Therefore, the mixed number is $4\frac{2}{3}$.

WRONG:

$\frac{14}{3}$; 14 divided by 3 equals 4, with a remainder of 3.

Therefore, the mixed number is $4\frac{3}{2}$.

Keep this in mind: You will use both of these conversion processes later in fraction problems with addition, subtraction, multiplication, and division.

BRAIN TICKLERS
Set # 1

Change the following funtastic fraction mixed numbers to improper fractions:

1. $2\frac{1}{4}$ 2. $3\frac{5}{6}$ 3. $4\frac{2}{5}$

4. $7\frac{1}{8}$ 5. $2\frac{5}{12}$ 6. $5\frac{4}{9}$

Now change these improper fractions to mixed numbers:

7. $\frac{11}{9}$ 8. $\frac{23}{4}$ 9. $\frac{16}{8}$

10. $\frac{55}{7}$ 11. $\frac{15}{2}$ 12. $\frac{37}{3}$

(Answers are on page 23.)

MULTIPLES

What is a multiple of a number?

A **multiple** of a number is the number multiplied by 1, 2, 3, 4, ...

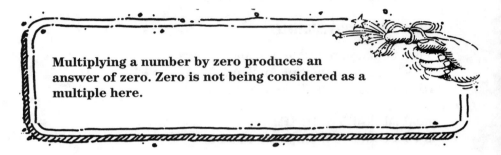

Multiplying a number by zero produces an answer of zero. Zero is not being considered as a multiple here.

What are the multiples of 6?

MARVELOUS MULTIPLES:

$6 \times 1 = 6, 6 \times 2 = 12, 6 \times 3 = 18, 6 \times 4 = 24$, and so on.

6, 12, 18, and 24 are multiples of 6.

MUTILATED MULTIPLES:

6, 12, 16, and 22 are multiples of 5.

These are not multiples of 5 because 5 doesn't divide into each one evenly.

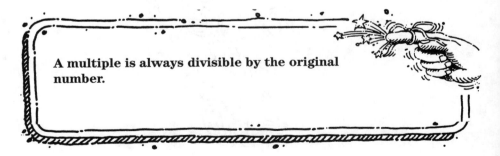

A multiple is always divisible by the original number.

What is a common multiple of two or more numbers?

A **common multiple** of two or more numbers is a multiple that is shared by two or more numbers. Two or more numbers may have more than one common multiple.

MORE MARVELOUS MULTIPLES:

6 is a common multiple of 2 and 6.
12 is a common multiple of 2 and 6.
10 is a common multiple of 2, 5, and 10.

MORE MUTILATED MULTIPLES:
Is 10 a common multiple of 4 and 6?

No, because 4 and 6 do not divide into 10 evenly.

Is 12 a common multiple of 5 and 10?

No, because 5 and 10 do not divide into 12 evenly.

What is an LCM? Is it a Little Corn Muffin?

No, it's a **Least Common Multiple (LCM)**.

A **Least Common Multiple (LCM)** is the lowest multiple that is shared by two or more numbers. It is a number into which all given numbers can divide evenly.

Exciting Example

What is the LCM of 6 and 10?

List multiples of 6.
$6 \times 1 = 6, 6 \times 2 = 12, 6 \times 3 = 18, 6 \times 4 = 24, \ldots$

6, 12, 18, 24, and 30 are multiples of 6.

List multiples of 10.
$10 \times 1 = 10, 10 \times 2 = 20, 10 \times 3 = 30, \ldots$

10, 20, and 30 are multiples of 30.

The LCM is 30. It is the **least common multiple** of 6 and 10. This is the lowest number that both 6 and 10 will divide into evenly.

• •

The LCM must be greater than or equal to one of your given numbers. The LCM will be used in fraction addition and subtraction problems to find a common denominator.

FACTORS

What is a factor of a number?

A **factor** of a number is a number that divides evenly into your given number.

What are the factors of 6?

FABULOUS FACTORS:
1, 2, 3, and 6 are the factors of 6 because they divide evenly into 6.

FRAUDULENT FACTORS:
4 and 5 are not factors of 6 because they do not divide evenly into 6.

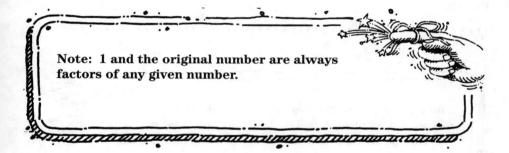

Note: 1 and the original number are always factors of any given number.

To determine factors, some divisibility tests
are useful:

Factor	Divisibility Test
2	If a number is even, it is divisible by 2. *Example:* 14 is an even number so it is divisible by 2.
5	If a number ends in 5, it is divisible by 5. *Example:* 25 ends in 5 so it is divisible by 5.
5, 10	If a number ends in 0, it is divisible by 5 and 10. *Example:* 30 ends in 0 so it is divisible by 5 and 10.
3	If the sum of the digits of a number is divisible by 3, the original number is divisible by 3. *Example:* In the number 513, add the digits 5 + 1 + 3 = 9. Since 9 divided by 3 equals 3, the original number, 513, is divisible by 3.
9	If the sum of the digits of a number is divisible by 9, the original number is divisible by 9. *Example:* In the number 783, add the digits 7 + 8 + 3 = 18. Since 18 divided by 9 equals 2, the original number, 783, is divisible by 9.
4	If the last two digits of a number are divisible by 4, then the original number is divisible by 4. *Example:* In the number, 3456, 56 is divisible by 4, so the original number is divisible by 4.

You may have to test numbers to see if they divide into your
given number evenly.

What is a common factor of two or more numbers?

A **common factor** of two or more numbers is a number that divides evenly into all of the numbers. Two numbers may have more than one common factor. A common factor can exist for more than two numbers.

FABULOUS FACTORS:
3 is a common factor of 3 and 6.
4 is a common factor of 4 and 12.
2 is a common factor of 4 and 12.
2 is a common factor of 4, 6, and 8.

FRAUDULENT FACTORS:
Is 4 a common factor of 6 and 8?

No, because 4 does not divide evenly into both 6 and 8.

Is 10 a common factor of 7 and 14?

No, because 10 does not divide evenly into both 7 and 14.

What is a GCF? Is it a Greasy Cheese Fry?

No, it's a **Greatest Common Factor (GCF)**.

A **Greatest Common Factor (GCF)** is the greatest factor that is shared by two or more numbers. It is a number that divides evenly into all of your given numbers.

Another Exciting Example

What is the GCF of 3 and 12?

List the factors of 3.

1, 3

List the factors of 12.

1, 2, 3, 4, 6, 12

The GCF is 3. It is the **greatest common factor** of 3 and 12.

This is the greatest number that divides evenly into both 3 and 12.

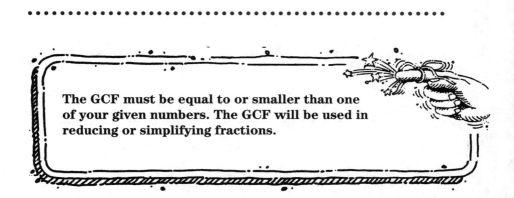

The GCF must be equal to or smaller than one of your given numbers. The GCF will be used in reducing or simplifying fractions.

Yet Another Exciting Example

What is the GCF of 36 and 45?

List the factors of 36.

1, 2, 3, 4, 6, 9, 12, 18, 36

List the factors of 45.

1, 3, 5, 9, 15, 45

The GCF is 9. It is the **greatest common factor** of 36 and 45.

This is the greatest number that divides evenly into both 36 and 45.

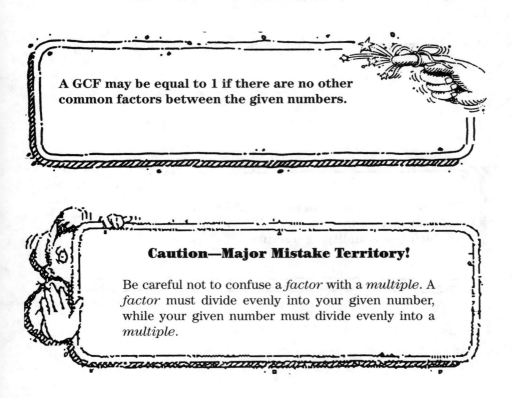

A GCF may be equal to 1 if there are no other common factors between the given numbers.

Caution—Major Mistake Territory!

Be careful not to confuse a *factor* with a *multiple*. A *factor* must divide evenly into your given number, while your given number must divide evenly into a *multiple*.

REDUCING FRACTIONS

How do you know if a fraction is in simplest form or lowest terms?

A fraction is in simplest form or lowest terms when no other number divides evenly into both the numerator and denominator except for 1, such as for $\frac{5}{6}$, $\frac{2}{3}$, or $\frac{1}{4}$.

Do you reduce fractions by putting them on a special diet?

Not really—reducing a fraction means changing the fraction to its simplest form or lowest terms. The reduced fraction is equivalent to the original fraction in a different form. To reduce a fraction to the simplest form or lowest terms, find the greatest number that divides evenly into both your numerator and denominator (the GCF). Then divide both the numerator and denominator by the GCF. The new numerator and denominator represent the reduced or simplified fraction.

For example, here's how to reduce $\frac{4}{6}$ to its lowest terms:

1. Find the GCF. In this problem, the GCF is 2.

2. Divide the numerator by the GCF. Divide 4 by 2, resulting in 2.

$$4 \div 2 = 2$$

3. Divide the denominator by the GCF. Divide 6 by 2, resulting in 3.

$$6 \div 2 = 3$$

The reduced or simplified fraction is $\frac{2}{3}$.

INTRIGUING INTRODUCTION

What do you do if you can't find a GCF?

If you can't find a GCF, use any number that divides evenly into both your numerator and denominator. Repeat as many times as necessary until your fraction is in simplest form. Remember that a fraction is in simplest form when no other number except 1 divides evenly into the numerator and denominator.

For example, here's how to reduce or simplify $\frac{24}{30}$ to its lowest terms:

1. Suppose you use 2 as your dividing number (instead of the GCF of 6).

2. Divide the numerator by 2. Divide 24 by 2, resulting in 12.

$$24 \div 2 = 12$$

3. Divide the denominator by 2. Divide 30 by 2, resulting in 15.

$$30 \div 2 = 15$$

Your new fraction is $\frac{12}{15}$. Can this fraction be reduced more? Yes it can—both the numerator and denominator can be divided by 3.

4. Divide the numerator by 3. Divide 12 by 3, resulting in 4.

$$12 \div 3 = 4$$

5. Divide the denominator by 3. Divide 15 by 3, resulting in 5.

$$15 \div 3 = 5$$

Your new fraction is $\frac{4}{5}$, which is in simplest form because no number other than 1 divides evenly into both 4 and 5.

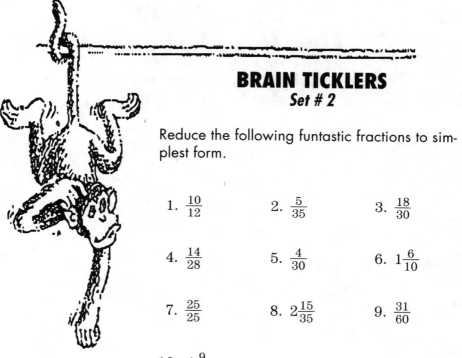

BRAIN TICKLERS
Set # 2

Reduce the following funtastic fractions to simplest form.

1. $\frac{10}{12}$

2. $\frac{5}{35}$

3. $\frac{18}{30}$

4. $\frac{14}{28}$

5. $\frac{4}{30}$

6. $1\frac{6}{10}$

7. $\frac{25}{25}$

8. $2\frac{15}{35}$

9. $\frac{31}{60}$

10. $4\frac{9}{36}$

(Answers are on page 25.)

BUILDING AN EQUIVALENT FRACTION

Do you build an equivalent fraction with bricks and mortar?

No, to build an equivalent fraction, use the multiplication property of 1. This property states that multiplying a number by 1 or any equivalent form of 1 does not change the value of the original number. A fraction can be changed into another equivalent fraction by multiplying by 1 (in any form). By doing this, the value of the original fraction is maintained.

An Exhilarating Example

Write $\frac{1}{3}$ as an equivalent fraction with a denominator of 12.

$$\frac{1}{3} = \frac{?}{12}$$

POPULAR PROCEDURE NUMBER 1:

1. Using the example above, divide 3 into 12, resulting in 4.

$$12 \div 3 = 4$$

2. Multiply 4 by the numerator 1, resulting in 4.

$$4 \times 1 = 4$$

3. Place the resulting 4 above the 12 as the new numerator. The equivalent fraction is $\frac{4}{12}$.

POPULAR PROCEDURE NUMBER 2:

1. Divide 3 into 12, resulting in 4.

$$12 \div 3 = 4$$

2. Use 4 to make a fraction equal to 1.

$$1 = \frac{4}{4}$$

3. Multiply $\frac{4}{4}$ by $\frac{1}{3}$ to obtain an equivalent fraction of $\frac{4}{12}$.

$$\frac{4}{4} \times \frac{1}{3} = \frac{4}{12}$$

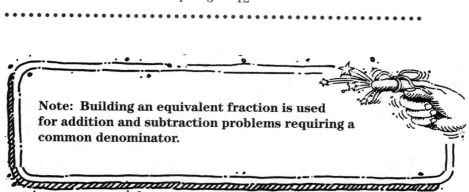

Note: Building an equivalent fraction is used for addition and subtraction problems requiring a common denominator.

Here's Another Exhilarating Example

This time, the original fraction has a numerator other than 1.

$$\frac{2}{5} = \frac{?}{25}$$

POPULAR PROCEDURE NUMBER 1:

1. Using the example above, divide 5 into 25, resulting in 5.

$$25 \div 5 = 5$$

2. Multiply 5 by the numerator 2, resulting in 10.

$$5 \times 2 = 10$$

3. Place the resulting 10 above the 25 as the new numerator. The equivalent fraction is $\frac{10}{25}$.

POPULAR PROCEDURE NUMBER 2:

1. Divide 5 into 25, resulting in 5.

$$25 \div 5 = 5$$

2. Use 5 to make a fraction equal to 1.

$$1 = \frac{5}{5}$$

3. Multiply $\frac{5}{5}$ by $\frac{2}{5}$ to obtain an equivalent fraction of $\frac{10}{25}$.

$$\frac{5}{5} \times \frac{2}{5} = \frac{10}{25}$$

For a problem such as $8 = \frac{?}{6}$, think of 8 as $\frac{8}{1}$ and proceed as before to fill in the ? with 48.

Here's a Slightly Different Exhilarating Example

What if you have $\frac{6}{9} = \frac{?}{15}$?

Because both of the previous methods can only be used when one denominator is divisible by the other with no remainder, you cannot use them here, as $15 \div 9$ results in a remainder. Instead, use a procedure called **cross multiplying**. Cross multiplying means to multiply the numerator of the first fraction by the denominator of the second fraction. This product will equal the denominator of the first fraction times the numerator of the second fraction.

1. Multiply 6×15 and $9 \times ?$ and set them equal to each other.

$$6 \times 15 = 9 \times ?$$

2. Multiply 6×15, resulting in 90.

$$90 = 9 \times ?$$

3. Divide 90 by 9.

$$90 \div 9 = 10$$

So, ? is equal to 10.

When you are finished building an equivalent fraction, you should be able to cross multiply your numbers and get two equivalent answers. For example,

$$\frac{1}{2} = \frac{3}{6}; \ 1 \times 6 = 2 \times 3$$

Because $6 = 6$, this serves as a check to prove your fractions are equivalent.

BRAIN TICKLERS
Set # 3

Convert the given funtastic fractions to equivalent funtastic fractions:

1. $\frac{2}{3} = \frac{?}{12}$ 2. $\frac{5}{6} = \frac{?}{18}$ 3. $\frac{1}{2} = \frac{?}{20}$

4. $\frac{2}{4} = \frac{?}{10}$ 5. $\frac{3}{4} = \frac{?}{20}$ 6. $\frac{8}{9} = \frac{?}{72}$

7. $\frac{1}{5} = \frac{?}{30}$ 8. $\frac{7}{9} = \frac{?}{63}$ 9. $10 = \frac{?}{4}$

10. $\frac{4}{11} = \frac{?}{33}$

(Answers are on page 28.)

BRAIN TICKLERS—THE ANSWERS

Set # 1, page 6

1. Multiply the denominator by the whole number and then add the numerator: $4 \times 2 + 1 = 9$. This 9 becomes the numerator, and the original denominator of 4 stays as the denominator. The improper fraction is $\frac{9}{4}$.

2. Multiply the denominator by the whole number and then add the numerator: $6 \times 3 + 5 = 23$. This 23 becomes the numerator, and the original denominator of 6 stays as the denominator. The improper fraction is $\frac{23}{6}$.

3. Multiply the denominator by the whole number and then add the numerator: $5 \times 4 + 2 = 22$. This 22 becomes the numerator, and the original denominator of 5 stays as the denominator. The improper fraction is $\frac{22}{5}$.

4. Multiply the denominator by the whole number and then add the numerator: $8 \times 7 + 1 = 57$. This 57 becomes the numerator, and the original denominator of 8 stays as the denominator. The improper fraction is $\frac{57}{8}$.

5. Multiply the denominator by the whole number and then add the numerator: $12 \times 2 + 5 = 29$. This 29 becomes the numerator, and the original denominator of 12 stays as the denominator. The improper fraction is $\frac{29}{12}$.

6. Multiply the denominator by the whole number and then add the numerator: $9 \times 5 + 4 = 49$. This 49 becomes the numerator, and the original denominator of 9 stays as the denominator. The improper fraction is $\frac{49}{9}$.

7. Divide 9 into 11: $9\overline{)11}$. You get a whole number of 1 with a remainder of 2. The remainder becomes the numerator of the fractional part and the divisor, 9, becomes the denominator. The mixed number is $1\frac{2}{9}$.

8. Divide 4 into 23: $4\overline{)23}$. You get a whole number of 5 with a remainder of 3. The remainder becomes the numerator of the fractional part and the divisor, 4, becomes the denominator. The mixed number is $5\frac{3}{4}$.

9. Divide 8 into 16: $8\overline{)16}$. You get a whole number of 2 with no remainder. The answer is 2.

10. Divide 7 into 55: $7\overline{)55}$. You get a whole number of 7 with a remainder of 6. The remainder becomes the numerator of the fractional part and the divisor, 7, becomes the denominator. The mixed number is $7\frac{6}{7}$.

11. Divide 2 into 15: $2\overline{)15}$. You get a whole number of 7 with a remainder of 1. The remainder becomes the numerator of the fractional part and the divisor, 2, becomes the denominator. The mixed number is $7\frac{1}{2}$.

12. Divide 3 into 37: $3\overline{)37}$. You get a whole number of 12 with a remainder of 1. The remainder becomes the numerator of the fractional part and the divisor, 3, becomes the denominator. The mixed number is $12\frac{1}{3}$.

Set # 2, page 17

In all problems below, if you find the GCF, your work will be quicker and easier. KEEP IN MIND: One and the original number are always factors of your given numbers.

1. To reduce $\frac{10}{12}$, look for a common factor that divides evenly into both the numerator and denominator. The factors of 10 are 1, 2, 5, 10 and the factors of 12 are 1, 2, 3, 4, 6, 12. The greatest common factor between 10 and 12 is 2. Divide the numerator of 10 by 2, $10 \div 2 = 5$, and the denominator of 12 by 2, $12 \div 2 = 6$. Make a fraction out of your result. The answer is $\frac{5}{6}$, which is in simplest form.

2. To reduce $\frac{5}{35}$, look for a common factor that divides evenly into both the numerator and denominator. The factors of 5 are 1 and 5 and the factors of 35 are 1, 5, 7, 35. The greatest common factor between 5 and 35 is 5. Divide the numerator of 5 by 5, $5 \div 5 = 1$, and the denominator of 35 by 5, $35 \div 5 = 7$. Make a fraction out of your result. The answer is $\frac{1}{7}$, which is in simplest form.

3. To reduce $\frac{18}{30}$, look for a common factor that divides evenly into both the numerator and denominator. The factors of 18 are 1, 2, 3, 6, 9, 18 and the factors of 30 are 1, 2, 3, 5, 6, 10, 15, 30. The greatest common factor between 18 and 30 is 6. Divide the numerator of 18 by 6, $18 \div 6 = 3$, and the denominator of 30 by 6, $30 \div 6 = 5$. Make a fraction out of your result. The answer is $\frac{3}{5}$, which is in simplest form.

4. To reduce $\frac{14}{28}$, look for a common factor that divides evenly into both the numerator and denominator. The factors of 14 are 1, 2, 7, 14 and the factors of 28 are 1, 2, 4, 7, 14, 28. The greatest common factor between 14 and 28 is 14. Divide the numerator of 14 by 14, $14 \div 14 = 1$, and the denominator of 28 by 14, $28 \div 14 = 2$. Make a fraction out of your result. The answer is $\frac{1}{2}$, which is in simplest form.

5. To reduce $\frac{4}{30}$, look for a common factor that divides evenly into both the numerator and denominator. The factors of 4 are 1, 2, 4 and the factors of 30 are 1, 2, 3, 5, 6, 10, 15, 30. The greatest common factor between 4 and 30 is 2. Divide the numerator of 4 by 2, $4 \div 2 = 2$, and the denominator of 30 by 2, $30 \div 2 = 15$. Make a fraction out of your result. The answer is $\frac{2}{15}$, which is in simplest form.

6. The whole number 1 stays the same. To reduce $\frac{6}{10}$, look for a common factor that divides evenly into both the numerator and denominator. The factors of 6 are 1, 2, 3, 6 and the factors of 10 are 1, 2, 5, 10. The greatest common factor between 6 and 10 is 2. Divide the numerator of 6 by 2, $6 \div 2 = 3$, and the denominator of 10 by 2, $10 \div 2 = 5$. Make a fraction out of your result. The answer is $\frac{3}{5}$, which is in simplest form, so your answer is the mixed number $1\frac{3}{5}$.

7. To reduce $\frac{25}{25}$, since both the numerator and denominator are the same, use 25 as your GCF. Divide 25 by 25, $25 \div 25 = 1$. Also, any number divided by itself is equal to 1.

8. The whole number 2 stays the same. To reduce $\frac{15}{35}$, look for a common factor that divides evenly into both the numerator and denominator. The factors of 15 are 1, 3, 5, 15 and the factors of 35 are 1, 5, 7, 35. The greatest common factor between 15 and 35 is 5. Divide the numerator of 15 by 5, $15 \div 5 = 3$, and the denominator of 35 by 5, $35 \div 5 = 7$. Make a fraction out of your result. The answer is $\frac{3}{7}$, which is in simplest form, so your answer is the mixed number $2\frac{3}{7}$.

9. Since the GCF of 31 and 60 is 1, the fraction $\frac{31}{60}$ is already in simplest form.

10. The whole number 4 stays the same. To reduce $\frac{9}{36}$, look for a common factor that divides evenly into both the numerator and denominator. The factors of 9 are 1, 3, 9 and the factors of 36 are 1, 2, 3, 4, 6, 9, 12, 18, 36. The greatest common factor between 9 and 36 is 9. Divide the numerator of 9 by 9, $9 \div 9 = 1$, and the denominator of 36 by 9, $36 \div 9 = 4$. Make a fraction out of your result. The fraction is $\frac{1}{4}$, which is in simplest form, so your answer is the mixed number $4\frac{1}{4}$.

Set # 3, page 22

1. Divide the denominator 3 into the denominator 12 to get 4.

$$\frac{4}{3\overline{)12}}$$

   Multiply this 4 by the numerator 2 to get 8: $4 \times 2 = 8$. Your new numerator is 8, so the equivalent fraction is equal to $\frac{8}{12}$.

2. Divide the denominator 6 into the denominator 18 to get 3.

$$\frac{3}{6\overline{)18}}$$

   Multiply this 3 by the numerator 5 to get 15: $3 \times 5 = 15$. Your new numerator is 15, so the equivalent fraction is equal to $\frac{15}{18}$.

3. Divide the denominator 2 into the denominator 20 to get 10.

$$\frac{10}{2\overline{)20}}$$

   Multiply this 10 by the numerator 1 to get 10: $10 \times 1 = 10$. Your new numerator is 10, so the equivalent fraction is equal to $\frac{10}{20}$.

4. Since you cannot divide the denominator 4 into the denominator 10 evenly, you must cross multiply 2×10 which equals 20 and $4 \times ?$ in the other direction. Setting these two diagonal products equal to each other, $20 = 4 \times ?$. Now to solve for ?, divide 4 into 20 to get 5.

$$\frac{5}{4\overline{)20}}$$

   Your new numerator is 5, so the equivalent fraction is equal to $\frac{5}{10}$.

5. Divide the denominator 4 into the denominator 20 to get 5.

$$4\overline{)20} \;\; 5$$

   Multiply this 5 by the numerator 3 to get 15: $5 \times 3 = 15$.
   Your new numerator is 15, so the equivalent fraction is equal
   to $\frac{15}{20}$.

6. Divide the denominator 9 into the denominator 72 to get 8.

$$9\overline{)72} \;\; 8$$

   Multiply this 8 by the numerator 8 to get 64: $8 \times 8 = 64$.
   Your new numerator is 64, so the equivalent fraction is equal
   to $\frac{64}{72}$.

7. Divide the denominator 5 into the denominator 30 to get 6.

$$5\overline{)30} \;\; 6$$

   Multiply this 6 by the numerator 1 to get 6: $1 \times 6 = 6$. Your
   new numerator is 6, so the equivalent fraction is equal to $\frac{6}{30}$.

8. Divide the denominator 9 into the denominator 63 to get 7.

$$9\overline{)63} \;\; 7$$

   Multiply this 7 by the numerator 7 to get 49: $7 \times 7 = 49$.
   Your new numerator is 49, so the equivalent fraction is equal
   to $\frac{49}{63}$.

9. Since 10 is a whole number, multiply this 10 by the denominator 4 to get 40: $10 \times 4 = 40$. Your new numerator is 40, so the equivalent fraction is equal to $\frac{40}{4}$.

10. Divide the denominator 11 into the denominator 33 to get 3.

$$11\overline{)33}$$ with quotient 3

Multiply this 3 by the numerator 4 to get 12: $3 \times 4 = 12$. Your new numerator is 12, so the equivalent fraction is equal to $\frac{12}{33}$.

CHAPTER TWO

Countless Comparisons

Multiplying by a fraction produces
an incredible shrinking action.

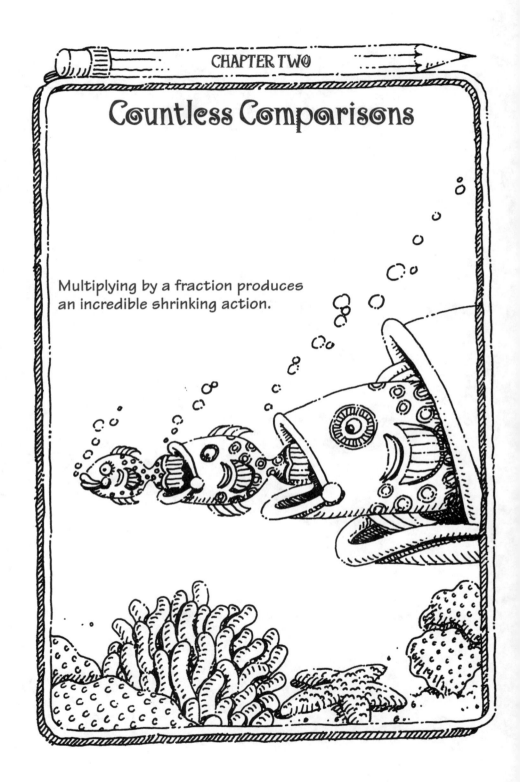

COMPARING TWO FRACTIONS

How do you compare the size of fractions?

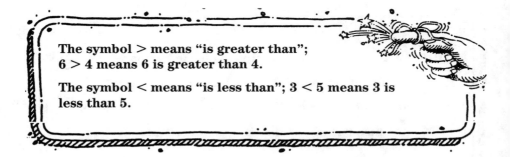

The symbol > means "is greater than";
6 > 4 means 6 is greater than 4.

The symbol < means "is less than"; 3 < 5 means 3 is
less than 5.

If the denominators are the same:
The one with the larger numerator is larger.

For example, to compare $\frac{2}{6}$ and $\frac{3}{6}$, notice that 3 is larger than
2. So, $\frac{3}{6}$ is larger than $\frac{2}{6}$, or $\frac{3}{6} > \frac{2}{6}$.

If the denominators are different:
Change the denominators into the lowest common denominator
(LCM) and see which numerator is larger.

COUNTLESS COMPARISONS

For example, to compare $\frac{1}{3}$ and $\frac{3}{4}$:

1. List multiples of 3: 3, 6, 9, 12 . . . . List multiples of 4: 4, 8, 12 . . . . The LCM of 3 and 4 is 12.

2. Change $\frac{1}{3}$ into $\frac{4}{12}$ by dividing 3 into 12 and multiplying the result by 1. Change $\frac{3}{4}$ into $\frac{9}{12}$ by dividing 4 into 12 and multiplying the result by 3.

3. Compare $\frac{4}{12}$ and $\frac{9}{12}$. Since 9 is larger than 4, $\frac{9}{12}$ is larger than $\frac{4}{12}$.

4. Thus $\frac{3}{4}$ is larger than $\frac{1}{3}$, or $\frac{3}{4} > \frac{1}{3}$.

SHORTCUT: For the same example, you can compare by cross multiplying. Multiply 4 by 1 and put the result above the 1. Multiply 3 by 3 and put the result above the 3: $^4\frac{1}{3} \times \frac{3}{4}^9$. **Since 9 is larger than 4, $\frac{3}{4}$ is larger than $\frac{1}{3}$.**

Caution—Major Mistake Territory!

This shortcut can only be used to compare two fractions. For three or more fractions you must find an LCM.

COMPARING THREE FRACTIONS

Put these three fractions in order from largest to smallest: $\frac{1}{2}, \frac{2}{3}, \frac{3}{4}$.

1. Find a common multiple of 2, 3, and 4. The LCM of 2, 3, and 4 is 12, so use 12 as the least common denominator.

2. $\frac{1}{2} = \frac{6}{12}$ because $12 \div 2 = 6$, and multiplying this 6 by the numerator of 1 gives 6 as the new numerator. $\frac{2}{3} = \frac{8}{12}$ because $12 \div 3 = 4$, and multiplying this 4 by the numerator of 2 gives 8 as the new numerator. $\frac{3}{4} = \frac{9}{1}$ because $12 \div 4 = 3$, and multiplying this 3 by the numerator of 3 gives 9 as the new numerator.

3. Since $\frac{1}{2} = \frac{6}{12}$, $\frac{2}{3} = \frac{8}{12}$, and $\frac{3}{4} = \frac{9}{12}$, the fractions in order from largest to smallest are $\frac{3}{4}, \frac{2}{3}, \frac{1}{2}$.

FLASH: If the numerators are all 1, the one with the smallest denominator is the largest fraction. For example, of $\frac{1}{2}, \frac{1}{3}$, and $\frac{1}{4}, \frac{1}{2}$ is the largest fraction.

COMPARING FRACTIONS, DECIMALS, AND PERCENTS

Now that you've compared the size of fractions, what if you have to compare a fraction to a decimal or to a percent?

You must change both to the same units, either fractions or decimals or percents.

To change a fraction to a decimal

Divide the denominator into the numerator. Add a decimal and at least one zero. Keep adding zeros, if necessary, to carry out the answer until it either comes out evenly, repeats, or can be rounded off to a given place.

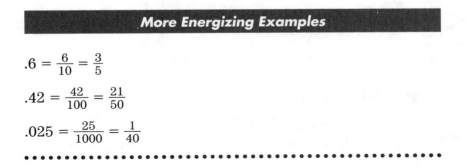

$\dfrac{1}{2} = 2\overline{)1.0}^{.5} = .5$

$\dfrac{5}{6} = 6\overline{)5.000}^{.833} = .8\overline{3}$ (repeating decimal has a line over it)

$\dfrac{2}{7} = 7\overline{)2.000}^{.285} = .285$ (round off to two decimal places). Since the digit in the third decimal place is 5 or more, round .285 to .29.

To change a decimal to a fraction:

Drop the decimal point. Make the remaining number the numerator of a fraction. The denominator of the fraction will be 10 if there is one decimal place, 100 if there are two decimal places, 1000 if there are three decimal places, etc. Then reduce the resulting fraction to simplest form.

$.6 = \dfrac{6}{10} = \dfrac{3}{5}$

$.42 = \dfrac{42}{100} = \dfrac{21}{50}$

$.025 = \dfrac{25}{1000} = \dfrac{1}{40}$

And Yet Another Energizing Example

Which is less? $\frac{1}{4}$ or 0.3?

GOOD:

$\frac{1}{4} = 4\overline{)1.00}^{0.25} = 0.25$ Compare 0.25 to 0.3. $.25 < .3$ because $.25 < .30$. Also, $\frac{25}{100} < \frac{30}{100}$. Therefore, $\frac{1}{4} < 0.3$.

ALSO GOOD:

Change 0.3 to $\frac{3}{10}$ since there is one place to the right of the decimal. Now compare $\frac{1}{4}$ to $\frac{3}{10}$. To compare $\frac{1}{4}$ to $\frac{3}{10}$ using the cross multiplication shortcut, multiply $10 \cdot 1$ and $4 \cdot 3$. Write your results over the appropriate fraction: $^{10}\frac{1}{4} \times \frac{3}{10}^{12}$. Since 10 is less than 12, $\frac{1}{4}$ is less than $\frac{3}{10}$. Therefore, $\frac{1}{4} < 0.3$.

NOT GOOD:

0.3 is less than $\frac{1}{4}$ because $3 < 4$.

To change a fraction to a percent:

Multiply the fraction by 100 (since percent is based on 100). Multiply 100 by the numerator and divide by the denominator. Simplify. Add a percent symbol.

To change a percent to a fraction:

Drop the percent symbol. Make the remaining number the numerator of a fraction. Make 100 the denominator. Reduce to simplify your fraction.

Example

Which is larger, $\frac{1}{5}$ or 40%?

PROPER:

Multiply to change $\frac{1}{5}$ to a percent: $\frac{1}{5} \cdot 100$. Multiply 100 by the numerator of 1 and divide by the denominator of 5. You will have $\frac{100}{5}$, which equals 20. Add the percent symbol to get 20%. Since 40% is larger than 20%, 40% is larger than $\frac{1}{5}$.

ALSO PROPER:

Change 40% to the fraction $\frac{40}{100}$. Reduce the fraction by dividing out the GCF of 20: $\frac{20}{20}$, $\frac{40 \div 20}{100 \div 20} = \frac{2}{5}$. Therefore, since $\frac{2}{5}$ is greater than $\frac{1}{5}$, 40% is larger than $\frac{1}{5}$.

IMPROPER:

$\frac{1}{5}$ is larger than 40% because $5 > 4$.

• •

If $\frac{2}{5}$ and $\frac{1}{5}$ didn't have the same denominator, you would have to rewrite both fractions with a common denominator or use the cross multiplication shortcut to determine which is larger.

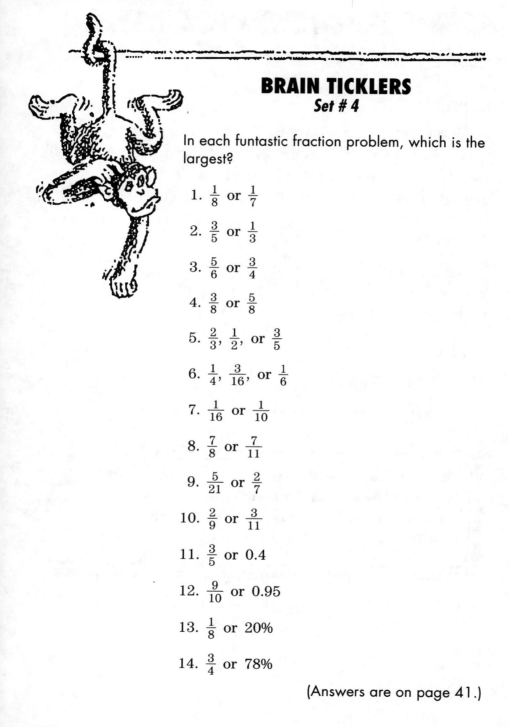

BRAIN TICKLERS
Set # 4

In each funtastic fraction problem, which is the largest?

1. $\frac{1}{8}$ or $\frac{1}{7}$

2. $\frac{3}{5}$ or $\frac{1}{3}$

3. $\frac{5}{6}$ or $\frac{3}{4}$

4. $\frac{3}{8}$ or $\frac{5}{8}$

5. $\frac{2}{3}$, $\frac{1}{2}$, or $\frac{3}{5}$

6. $\frac{1}{4}$, $\frac{3}{16}$, or $\frac{1}{6}$

7. $\frac{1}{16}$ or $\frac{1}{10}$

8. $\frac{7}{8}$ or $\frac{7}{11}$

9. $\frac{5}{21}$ or $\frac{2}{7}$

10. $\frac{2}{9}$ or $\frac{3}{11}$

11. $\frac{3}{5}$ or 0.4

12. $\frac{9}{10}$ or 0.95

13. $\frac{1}{8}$ or 20%

14. $\frac{3}{4}$ or 78%

(Answers are on page 41.)

BRAIN TICKLERS—THE ANSWERS

Set # 4, page 40

1. Since the numerators are both 1, the fraction with the smaller denominator is the larger fraction. Therefore, $\frac{1}{7}$ is larger than $\frac{1}{8}$.

2. To compare $\frac{3}{5}$ and $\frac{1}{3}$, use the cross multiplication shortcut. Multiply 3 by 3 and put the result above the 3 on the left. Then multiply 5 by 1 and put the result above the 1 on the right.

$$^9\frac{3}{5} \times \frac{1}{3}^5$$

Since 9 is larger than 5, the fraction under the 9, $\frac{3}{5}$, is larger than the fraction under the 5, $\frac{1}{3}$. Therefore, $\frac{3}{5}$ is larger than $\frac{1}{3}$.

3. To compare $\frac{5}{6}$ and $\frac{3}{4}$, use the cross multiplication shortcut. Multiply 4 by 5 and put the result above the 5 on the left. Then multiply 6 by 3 and put the result above the 3 on the right.

$$^{20}\frac{5}{6} \times \frac{3}{4}^{18}$$

Since 20 is larger than 18, the fraction on the left, $\frac{5}{6}$ is larger than the fraction on the right. Therefore, $\frac{5}{6}$ is larger than $\frac{3}{4}$.

4. Since $\frac{3}{8}$ and $\frac{5}{8}$ have the same denominator, you can compare the size of the numerators to see which is larger. Since 5 is larger than 3, the fraction $\frac{5}{8}$ is larger than $\frac{3}{8}$.

5. For three fractions, find a common denominator so that the size of the numerators can be compared. To find a common denominator, list the multiples of all three denominators. Multiples of 3 are 3, 6, 9, 12, 15, 18, 21, 24, 27, 30 . . . ; multiples of 2 are 2, 4, 6, 8, 10, 12, 14, 16, 18, 20, 22, 24, 26, 28, 30 . . . ; and multiples of 5 are 5, 10, 15, 20, 25, 30 . . . . The common denominator is the LCM, which is equal to 30. Rewrite $\frac{2}{3}$ with the common denominator of 30, $\frac{2}{3} = \frac{?}{30}$. To do this divide 3 into 30 to get 10, and multiply 10 by 2 to fill in the ? with 20. So, $\frac{2}{3} = \frac{20}{30}$. Similarly, $\frac{1}{2} = \frac{15}{30}$, and $\frac{3}{5} = \frac{18}{30}$. With 30 as the common denominator, you can see that $\frac{20}{30}$ is the largest, then $\frac{18}{30}$, then $\frac{15}{30}$ in decreasing order. Therefore, the order of the original fractions from largest to smallest is $\frac{2}{3}, \frac{3}{5}, \frac{1}{2}$.

6. For three fractions, find a common denominator and rewrite each fraction with this common denominator so that the size of the numerators can be compared. To find a common denominator, list the multiples of all three denominators. Multiples of 4 are 4, 8, 12, 16, 20, 24, 28, 32, 36, 40, 44, 48 . . . ; multiples of 16 are 16, 32, 48, 64 . . . ; and multiples of 6 are 6, 12, 18, 24, 30, 36, 42, 48, 54 . . . . The LCM is 48, which can be used as the common denominator. Rewrite $\frac{1}{4}$ with the common denominator of 48, $\frac{1}{4} = \frac{?}{48}$. To do this, divide 4 into 48, to get 12, and multiply 12 by 1 to fill in the ? with 12. So, $\frac{1}{4} = \frac{12}{48}$. Similarly, $\frac{3}{16} = \frac{9}{48}$, and $\frac{1}{6} = \frac{8}{48}$. With 48 as the denominator, you can see that $\frac{12}{48}$ is the largest, then $\frac{9}{48}$, then $\frac{8}{48}$ in decreasing order. Therefore, the order of the original fractions from largest to smallest is $\frac{1}{4}, \frac{3}{16}$, and $\frac{1}{6}$.

7. Since the numerators are both 1, the fraction with the smaller denominator is the larger fraction. Therefore, $\frac{1}{10}$ is larger than $\frac{1}{16}$.

8. To compare $\frac{7}{8}$ and $\frac{7}{11}$, use the cross multiplication shortcut. Multiply 11 by 7 and put the result above the 7 on the left. Then multiply 8 by 7 and put your answer above the 7 on the right.

$$^{77}\,\frac{7}{8} \times \frac{7}{11}\,^{56}$$

Since 77 is larger than 56, the fraction on the left, $\frac{7}{8}$, is larger than the fraction on the right. Therefore, $\frac{7}{8}$ is larger than $\frac{7}{11}$.

9. To compare $\frac{5}{21}$ and $\frac{2}{7}$, use the cross multiplication shortcut. Multiply 7 by 5 and put the result above the 5 on the left. Then multiply 21 by 2 and put the result above the 2 on the right.

$$^{35}\,\frac{5}{21} \times \frac{2}{7}\,^{42}$$

Since 42 is larger than 35, the fraction on the right, $\frac{2}{7}$, is larger than the fraction on the left, $\frac{5}{21}$. Therefore, $\frac{2}{7}$ is larger than $\frac{5}{21}$.

10. To compare $\frac{2}{9}$ and $\frac{3}{11}$, use the cross multiplication shortcut. Multiply 11 by 2 and put the result above the 2 on the left. Then multiply 9 by 3 and put the result above the 3 on the right.

$$^{22}\,\frac{2}{9} \times \frac{3}{11}\,^{27}$$

Since 27 is larger than 22, the fraction on the right, $\frac{3}{11}$, is larger than the fraction on the left. Therefore, $\frac{3}{11}$ is larger than $\frac{2}{9}$.

11. Change $\frac{3}{5}$ into a decimal by dividing 5 into 3: $5\overline{)3.0}$ with quotient 0.6.

So, 0.6 is equivalent to $\frac{3}{5}$. Therefore, since $0.6 > 0.4$, $\frac{3}{5}$ is larger than 0.4.

12. Change $\frac{9}{10}$ into a decimal by dividing 10 into 9: $10\overline{)9.0}^{\,0.9}$.

    So, 0.9 is equivalent to $\frac{9}{10}$. Since 0.95 > 0.9, 0.95 is

    larger than $\frac{9}{10}$.

13. Multiply to change $\frac{1}{8}$ to a percent: $\frac{1}{8} \cdot 100$. Multiply 100 by
    the numerator of 1 and divide by the denominator of 8 to
    get $\frac{100 \cdot 1}{8}$ and simplify to $\frac{25}{2}$. Changing $\frac{25}{2}$ to a mixed
    number produces $12\frac{1}{2}$. Add the percent symbol to get an
    answer of $12\frac{1}{2}\%$. Therefore, since 20% > $12\frac{1}{2}\%$, 20% is
    larger than $\frac{1}{8}$.

14. Multiply to change $\frac{3}{4}$ to a percent: $\frac{3}{4} \cdot 100$. Multiply 100 by
    the numerator of 3 and divide by the denominator of 4 to
    get $\frac{300}{4}$; which equals 75. Add the percent symbol to get an
    answer of 75%. Therefore, since 78% > 75%, 78% is larger
    than $\frac{3}{4}$.

CHAPTER THREE

Magnificent Multiplication

Multiplication can be represented in three ways: ×, ·, or ( ). Unlike whole numbers, when you multiply two proper fractions, you will get a smaller resulting fraction than either of your original fractions.

TWO PROPER FRACTIONS

What happens when you multiply two proper fractions, where there are no common factors between the numerators and the denominators?

For example, in $\frac{1}{2} \times \frac{3}{4}$, there are no common factors between the numerators and denominators. No numbers are the same, and there is no number that divides evenly into both a number in the numerators and a number in the denominators. So, you simply multiply the numerators and the denominators to get $\frac{1}{2} \times \frac{3}{4} = \frac{3}{8}$.

What do you do with two proper fractions with the same number in one of the numerators and one of the denominators?

To shorten the multiplication process and simplify your result, use a method called **canceling**. To cancel your fractions, find a common factor that is the same as or divides evenly into both a number in the numerators and a number in the denominators. For example, multiply $\frac{1}{2} \cdot \frac{2}{3}$.

1. Since 2 appears in both a numerator and a denominator, the fractions can be canceled by dividing both the numerator and the denominator by 2.

2. In $\frac{1}{1\cancel{2}} \cdot \frac{\cancel{2}^1}{3}$, each of the 2s is replaced by 1s. The problem now looks like this: $\frac{1}{1} \cdot \frac{1}{3}$.

3. Now multiply the numerators and the denominators, resulting in $\frac{1}{3}$.

How do you multiply two fractions with a common factor between the numerators and denominators?

For example, multiply $\frac{1}{2} \cdot \frac{4}{5}$.

1. Since 2 is a common factor to both 2 and 4, divide 2 into each of these numbers.

$$4 \div 2 = 2$$
$$\frac{1}{1\cancel{2}} \cdot \frac{\cancel{4}^2}{5}$$
$$2 \div 2 = 1$$

2. The problem now looks like this: $\frac{1}{1} \cdot \frac{2}{5}$.

3. Multiply the numerators and the denominators, resulting in $\frac{2}{5}$.

WHOLE NUMBERS WITH PROPER FRACTIONS

What if you want to multiply a whole number by a proper fraction?

Nice Example

Multiply $4 \cdot \frac{3}{8}$.

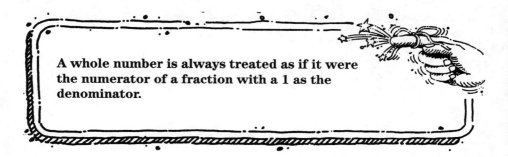

A whole number is always treated as if it were the numerator of a fraction with a 1 as the denominator.

1. Treat your whole number 4 as if it were the numerator in $\frac{4}{1}$.

2. Use canceling as in the previous problem to divide 4 into both 4 and 8.

$$8 \div 4 = 2$$
$$4 \div 4 = 1$$

$$\overset{1}{4} \cdot \frac{3}{\underset{2}{8}}$$

3. The problem now looks like this: $1 \cdot \frac{3}{2}$.

4. Multiply the numerators and the denominators, resulting in $\frac{3}{2}$.

$$\frac{1}{1} \cdot \frac{3}{2} = \frac{3}{2}$$

5. Simplify the improper fraction by dividing 3 by 2, resulting in $1\frac{1}{2}$.

$$3 \div 2 = 1\frac{1}{2}$$

• •

MIXED NUMBERS

What if you have two mixed numbers?

Nifty Example

Multiply $3\frac{3}{7} \times 2\frac{1}{3}$.

1. Change each of the mixed numbers to improper fractions.

$$3\frac{3}{7} = \frac{24}{7}$$

$$2\frac{1}{3} = \frac{7}{3}$$

The new representation is $\frac{24}{7} \times \frac{7}{3}$.

2. Cancel each of the 7s replacing them by 1s.

$$7 \div 7 = 1 \qquad \frac{24}{{}_{1}7} \cdot \frac{7^{1}}{3}$$

The problem now looks like this: $\frac{24}{1} \times \frac{1}{3}$.

3. Cancel both the 3 and 24 by dividing by 3.

$$3 \div 3 = 1$$
$$24 \div 3 = 8 \qquad \frac{{}^{8}24}{1} \cdot \frac{1}{{}_{1}\cancel{3}}$$

The problem now looks like this: $\frac{8}{1} \times \frac{1}{1}$.

4. $\frac{8 \times 1}{1 \times 1} = \frac{8}{1} = 8$

The result is 8.

• •

Not So Nifty Example

$$3\frac{3}{7} \times 2\frac{1}{3} = 6\frac{3}{21}$$

• •

Caution—Major Mistake Territory!

Don't multiply the two whole numbers and then the two fractions.

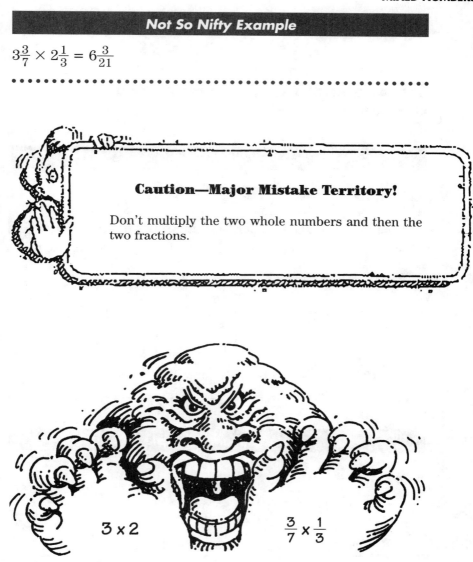

3×2 $\frac{3}{7} \times \frac{1}{3}$

WHOLE NUMBERS WITH MIXED NUMBERS

What about a whole number and a mixed number?

Pleasant Example

Multiply $5 \cdot 8\frac{4}{5}$.

1. Change the mixed number to an improper fraction, $8\frac{4}{5} = \frac{44}{5}$, and the whole number to $\frac{5}{1}$. The problem now looks like this: $\frac{5}{1} \cdot \frac{44}{5}$.

2. Cancel the 5s.

$$5 \div 5 = 1 \qquad \frac{{}^1\cancel{5}}{1} \cdot \frac{44}{\cancel{5}_1}$$

The result is $\frac{44}{1}$, or 44.

● ●

Unpleasant Example

$5 \cdot 8\frac{4}{5} = 40\frac{4}{5}$

● ●

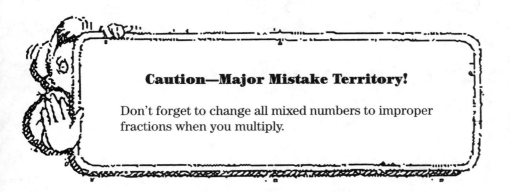

Caution—Major Mistake Territory!

Don't forget to change all mixed numbers to improper fractions when you multiply.

THREE OR MORE FRACTIONS

Is the procedure the same for three or more fractions?

Sweet Example

Multiply $2\frac{1}{2} \cdot \frac{1}{4} \cdot \frac{3}{5}$.

1. Change mixed numbers to improper fractions.

$$2\frac{1}{2} = \frac{5}{2}$$

   The problem now looks like this: $\frac{5}{2} \cdot \frac{1}{4} \cdot \frac{3}{5}$.

2. You may cancel *any numerator* with *any denominator*, regardless of location. In this case, cancel the 5s.

$$5 \div 5 = 1 \qquad \frac{{}^{1}\cancel{5}}{2} \cdot \frac{1}{4} \cdot \frac{3}{\cancel{5}_{1}}$$

   The problem now looks like this: $\frac{1}{2} \cdot \frac{1}{4} \cdot \frac{3}{1}$.

3. Now multiply the numerators and the denominators.

$$\frac{1 \cdot 1 \cdot 3 = 3}{2 \cdot 4 \cdot 1 = 8}$$

4. The result is $\frac{3}{8}$.

• •

Sour Example

$2\frac{1}{2} \cdot \frac{1}{4} \cdot \frac{3}{5} = 2\frac{3}{40}$ by just multiplying the numerators, $1 \cdot 1 \cdot 3 = 3$, and the denominators $2 \cdot 4 \cdot 5 = 40$, to make $2\frac{3}{40}$.

Fractions can be canceled vertically as well as diagonally. A numerator and denominator in the same fraction can be canceled with each other. For example, in $\frac{3}{8} \cdot \frac{5}{20}$, you can cancel the 5 with the 20 in $\frac{5}{20}$ because $5 \div 5 = 1$ and $20 \div 5 = 4$.

$$\frac{3}{8} \cdot \frac{\cancel{5}^{1}}{\cancel{20}_{4}}$$

The problem now reads $\frac{3}{8} \cdot \frac{1}{4}$. Multiply $\frac{3 \cdot 1 = 3}{8 \cdot 4 = 32}$ to get $\frac{3}{32}$.

Caution—Major Mistake Territory!

Never cancel two numbers in the numerator or two numbers in the denominator. Fractions can never be canceled horizontally. One number must be from the numerator and one number must be from the denominator.

Multiplying a whole number or a fraction by 0 produces a result of 0.

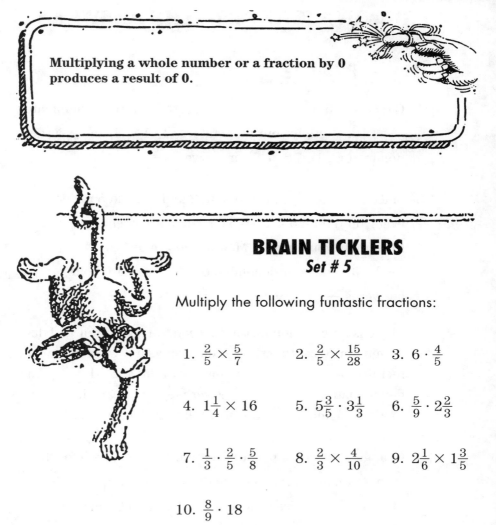

BRAIN TICKLERS
Set # 5

Multiply the following funtastic fractions:

1. $\frac{2}{5} \times \frac{5}{7}$

2. $\frac{2}{5} \times \frac{15}{28}$

3. $6 \cdot \frac{4}{5}$

4. $1\frac{1}{4} \times 16$

5. $5\frac{3}{5} \cdot 3\frac{1}{3}$

6. $\frac{5}{9} \cdot 2\frac{2}{3}$

7. $\frac{1}{3} \cdot \frac{2}{5} \cdot \frac{5}{8}$

8. $\frac{2}{3} \times \frac{4}{10}$

9. $2\frac{1}{6} \times 1\frac{3}{5}$

10. $\frac{8}{9} \cdot 18$

(Answers are on page 56.)

BRAIN TICKLERS—THE ANSWERS

Set # 5, page 55

1. Cancel the 5 in the numerator with the 5 in the denominator, $\frac{2}{1\,\cancel{5}} \cdot \frac{\cancel{5}^1}{7}$. Multiply your numerators, $2 \cdot 1 = 2$, and your denominators, $1 \cdot 7 = 7$. Your answer is $\frac{2}{7}$.

2. Since 2 divides evenly into both 2 and 28, and 5 divides evenly into both 5 and 15, you can use canceling, $\frac{^1\cancel{2}}{_1\cancel{5}} \cdot \frac{\cancel{15}}{\cancel{28}_{14}}$, to simplify your answer. Multiply your numerators, $1 \cdot 3 = 3$, and your denominators, $1 \cdot 14 = 14$. Your answer is $\frac{3}{14}$.

3. There is no common factor between the numerator and denominator. Treat the whole number 6 as if it were in the numerator because $6 = \frac{6}{1}$ and multiply it by 4, $6 \cdot 4 = 24$. The denominator is 5, so your answer is $\frac{24}{5}$. You can change this to a mixed number by dividing 5 into 24 to equal $4\frac{4}{5}$.

4. Change $1\frac{1}{4}$ to an improper fraction. $4 \cdot 1 + 1 = 5$ for the numerator, and 4 remains the denominator. So, $1\frac{1}{4} = \frac{5}{4}$. Multiply $\frac{5}{4} \cdot 16$. Use canceling for 4 and 16 because 4 divides evenly into both, $\frac{5}{_1\cancel{4}} \cdot \cancel{16}^4$. Multiply the remaining numbers to get an answer of 20.

5. Change $5\frac{3}{5}$ to an improper fraction: $5 \cdot 5 + 3 = 28$ for the numerator and 5 remains the denominator. So, $5\frac{3}{5} = \frac{28}{5}$. Also change $3\frac{1}{3}$ to an improper fraction: $3 \cdot 3 + 1 = 10$ for the numerator, and 3 remains the denominator. So, $3\frac{1}{3} = \frac{10}{3}$. Multiply $\frac{28}{5} \cdot \frac{10}{3}$. Use canceling to divide 5 into both 5 and 10, $\frac{28}{1\cancel{5}} \cdot \frac{\cancel{10}^{2}}{3}$. Multiply the remaining numbers to get an answer of $\frac{56}{3}$. Change this to a mixed number by dividing 3 into 56 to get $18\frac{2}{3}$.

6. Change $2\frac{2}{3}$ to an improper fraction: $3 \cdot 2 + 2 = 8$ for the numerator, and 3 remains the denominator. So, $2\frac{2}{3} = \frac{8}{3}$. Multiply $\frac{5}{9} \cdot \frac{8}{3}$. There are no common factors between the numerators and the denominators, so multiply $5 \cdot 8$ in the numerator and $9 \cdot 3$ in the denominator to get an answer of $\frac{40}{27}$. Change to a mixed number by dividing 27 into 40 to get $1\frac{13}{27}$.

7. Here, the 5 in the second denominator and the 5 in the third numerator can be canceled, $\frac{1}{3} \cdot \frac{2}{1\cancel{5}} \cdot \frac{\cancel{5}^{1}}{8}$. Next, use canceling to divide 2 into itself in the second numerator and into 8 in the third denominator, $\frac{1}{3} \cdot \frac{\cancel{2}^{1}}{1} \cdot \frac{1}{\cancel{8}_{4}}$. Multiply the remaining numbers in the numerator to get 1 and in the denominator to get 12. Your resulting fraction is $\frac{1}{12}$.

8. In $\frac{2}{3} \cdot \frac{4}{10}$, you can use canceling in one of two ways:
   (1) Divide 2 into itself and into 10 on a diagonal, $\frac{{}^{1}\cancel{2}}{3} \cdot \frac{4}{\cancel{10}\,5}$.
   Now multiply your numerators, $1 \cdot 4 = 4$, and your denominators, $3 \cdot 5 = 15$. Your answer is $\frac{4}{15}$. (2) In the second
   method, you can cancel in an up/down fashion in the second
   fraction. Divide 2 into 4 and 2 into 10, $\frac{2}{3} \cdot \frac{\cancel{4}\,^{2}}{\cancel{10}\,5}$. Now multiply your numerators, $2 \cdot 2 = 4$, and your denominators, $3 \cdot 5 = 15$. Your answer is $\frac{4}{15}$.

9. Change $2\frac{1}{6}$ to an improper fraction. $6 \cdot 2 + 1 = 13$ for the
   numerator, and 6 remains the denominator. So, $2\frac{1}{6} = \frac{13}{6}$.
   Now, change $1\frac{3}{5}$ to an improper fraction. $5 \cdot 1 + 3 = 8$
   for the numerator, and 5 remains the denominator. So,
   $1\frac{3}{5} = \frac{8}{5}$. You now have $\frac{13}{6} \cdot \frac{8}{5}$. Use canceling to divide 2 into both 6 and 8, $\frac{13}{{}_{3}\cancel{6}} \cdot \frac{\cancel{8}\,^{4}}{5}$. Your product now
   looks like $\frac{13}{3} \cdot \frac{4}{5}$. Multiply your numerators, $13 \cdot 4 = 52$,
   and your denominators $3 \cdot 5 = 15$. Your fraction is $\frac{52}{15}$.
   Divide 15 into 52 to get $3\frac{7}{15}$.

10. Use canceling to divide 9 into both 9 and 18, $\frac{8}{{}_{1}\cancel{9}} \cdot \cancel{18}\,^{2}$.
    Multiply your remaining numbers to give an answer of 16.

CHAPTER FOUR

Dazzling Division in a Fraction of the Time

When you divide whole numbers
Your result is quite small,
But we fractions when divided,
Yield a larger result after all.

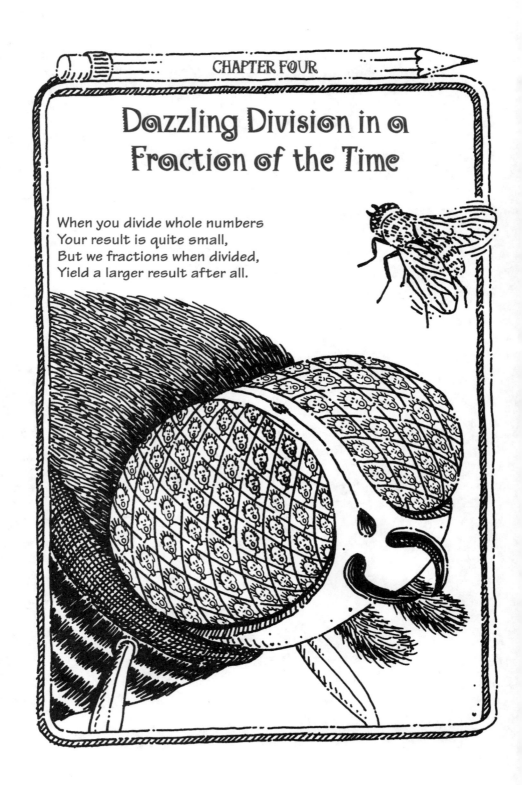

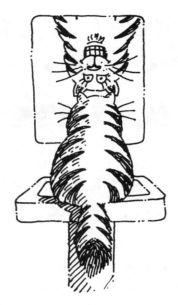

A **reciprocal** of a fraction is a fraction where the numerator and denominator are reversed from the original.

EXAMPLE:

$\frac{2}{3}$ and $\frac{3}{2}$ are reciprocals

EXAMPLE:

$\frac{1}{4}$ and 4 are reciprocals

A fraction multiplied by its reciprocal equals 1. To divide two fractions, multiply the first fraction by the reciprocal of the second fraction. Use the rules for multiplication of fractions.

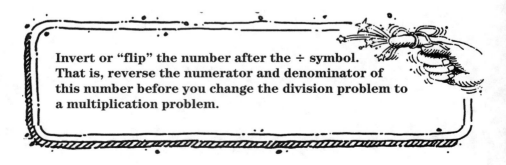

Invert or "flip" the number after the ÷ symbol. That is, reverse the numerator and denominator of this number before you change the division problem to a multiplication problem.

TWO PROPER FRACTIONS

How do I divide two proper fractions?

Superb Example

Find $\frac{1}{2} \div \frac{1}{3}$.

1. Change the $\frac{1}{3}$ to $\frac{3}{1}$, because $\frac{3}{1}$ is the reciprocal of $\frac{1}{3}$.

2. Change the division symbol ÷ to a multiplication symbol ×, ·, or ( ).

3. Use multiplication rules and simplify your answer. The problem now looks like $\frac{1}{2} \cdot \frac{3}{1} = \frac{3}{2}$, which equals $1\frac{1}{2}$.

• •

Silly Example

$\frac{1}{2} \div \frac{1}{3} = \frac{1}{6}$

You can't multiply the two numerators and the denominators in a division problem.

• •

TWO MIXED NUMBERS

What about two mixed numbers?

Find $3\frac{1}{2} \div 2\frac{3}{4}$.

1. Change **mixed numbers** to **improper fractions**.

$$3\frac{1}{2} = \frac{7}{2} \qquad 2\frac{3}{4} = \frac{11}{4}$$

2. The problem now looks like $\frac{7}{2} \div \frac{11}{4}$.

3. Change the $\frac{11}{4}$ to $\frac{4}{11}$ and the division symbol to multiplication.

4. The problem now looks like $\frac{7}{2} \times \frac{4}{11}$.

5. Cancel the 2 and 4 because $4 \div 2 = 2$ and $2 \div 2 = 1$:
$$\frac{7}{{}_1\cancel{2}} \times \frac{\cancel{4}^{\,2}}{11}.$$

6. The problem now looks like $\frac{7}{1} \times \frac{2}{11}$.

7. Use multiplication rules to simplify your answer. $\frac{7 \times 2 = 14}{1 \times 11 = 11}$.
Your answer is $\frac{14}{11}$, which equals $1\frac{3}{11}$.

DORKY:

$$3\frac{1}{2} \div 2\frac{3}{4} = \frac{7}{2} \times \frac{11}{4} = \frac{77}{8} = 9\frac{5}{8}$$

DORKIER:

$$3\frac{1}{2} \div 2\frac{3}{4} = 3\frac{1}{2} \times 2\frac{3}{4} = 6\frac{3}{8}$$

DORKIEST:

$$3\frac{1}{2} \div 2\frac{3}{4} = 3\frac{2}{1} \times 2\frac{4}{3} = 6\frac{8}{3}$$

• •

WHOLE NUMBER AND
PROPER FRACTION

What about a whole number and a proper fraction?

Awesome Example

Find $6 \div \frac{2}{7}$.

1. Invert the $\frac{2}{7}$ to $\frac{7}{2}$, because $\frac{7}{2}$ is the reciprocal of $\frac{2}{7}$.

Always invert the number *after* the ÷ symbol.

---

2. Cancel the 2 and the 6 by dividing 2 into each number.

$$6 \div 2 = 3 \qquad \overset{3}{\cancel{6}} \times \frac{7}{\underset{1}{\cancel{2}}}$$
$$2 \div 2 = 1$$

3. The problem now looks like $3 \times \frac{7}{1}$. The answer is 21.

AWFUL:

$$6 \div \frac{2}{7} = \frac{1}{6} \times \frac{2}{7} = \frac{2}{42} = \frac{1}{21}$$

MIXED NUMBER AND WHOLE NUMBER

How do I handle a mixed number and a whole number?

Get a grip on it of course.

Fabulous Example

Find $1\frac{2}{7} \div 6$.

1. Change the $1\frac{2}{7}$ to the improper fraction $\frac{9}{7}$.

2. Think of 6 as $\frac{6}{1}$ and invert it to $\frac{1}{6}$.

3. The problem now looks like $\frac{9}{7} \times \frac{1}{6}$.

4. Use canceling (by dividing 3 into both 6 and 9).

$$6 \div 3 = 2 \qquad \frac{\overset{3}{\cancel{9}}}{7} \times \frac{1}{\underset{2}{\cancel{6}}}$$

$$9 \div 3 = 3$$

5. The problem now looks like $\frac{3}{7} \times \frac{1}{2}$.

6. Use multiplication rules to simplify your answer:

$$\frac{3 \times 1 = 3}{7 \times 2 = 14}.$$

Your answer is $\frac{3}{14}$.

FRAUDULENT:

$$1\frac{2}{7} \div 6 = \frac{9}{7} \cdot 6 = \frac{54}{7} = 7\frac{5}{7}$$

• •

Caution—Major Mistake Territory!

0 in the numerator of a fraction makes the fraction equal to 0.

0 in the denominator of a fraction makes the fraction "undefined" because you can't divide by 0.

EXAMPLE:

$\frac{0}{6} = 0$ because $6\overline{)0}^{\,0}$ and $0 \cdot 6 = 0$ to check.

EXAMPLE:

$\frac{6}{0} =$ undefined because $0\overline{)6}^{\,?}$ and there is no number that can be multiplied by 0 to give you 6.

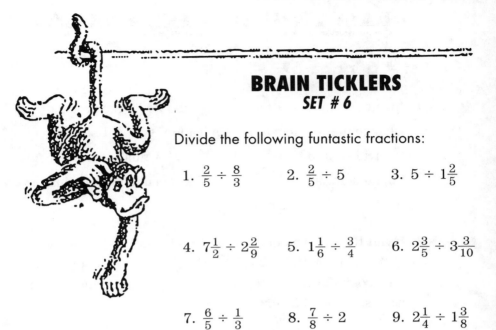

BRAIN TICKLERS
SET # 6

Divide the following funtastic fractions:

1. $\frac{2}{5} \div \frac{8}{3}$ 2. $\frac{2}{5} \div 5$ 3. $5 \div 1\frac{2}{5}$

4. $7\frac{1}{2} \div 2\frac{2}{9}$ 5. $1\frac{1}{6} \div \frac{3}{4}$ 6. $2\frac{3}{5} \div 3\frac{3}{10}$

7. $\frac{6}{5} \div \frac{1}{3}$ 8. $\frac{7}{8} \div 2$ 9. $2\frac{1}{4} \div 1\frac{3}{8}$

10. $\frac{7}{3} \div 2\frac{4}{5}$

(Answers are on page 68.)

67

BRAIN TICKLERS—THE ANSWERS

Set # 6, page 67

1. Invert the number after the $\div$ symbol to its reciprocal of $\frac{3}{8}$.
   Change the problem to multiplication, $\frac{2}{5} \times \frac{3}{8}$. Use canceling
   to divide 2 into both 2 and 8, $\frac{1\,\cancel{2}}{5} \times \frac{3}{\cancel{8}\,4}$. Multiply the remaining numerators, $1 \cdot 3 = 3$, and the remaining denominators,
   $5 \cdot 4 = 20$. Your answer is $\frac{3}{20}$.

2. Invert the whole number 5 to its reciprocal of $\frac{1}{5}$. Change the
   problem to multiplication, $\frac{2}{5} \times \frac{1}{5}$. There are no common factors, so multiply the numerators, $2 \cdot 1 = 2$, and the denominators, $5 \cdot 5 = 25$. Your answer is $\frac{2}{25}$.

3. Change $1\frac{2}{5}$ to an improper fraction, $5 \cdot 1 + 2 = 7$ is the
   numerator, and 5 remains as the denominator. So, $1\frac{2}{5} = \frac{7}{5}$.
   The problem now reads $5 \div \frac{7}{5}$. Invert $\frac{7}{5}$ to its reciprocal of $\frac{5}{7}$
   and change the problem to multiplication, $5 \times \frac{5}{7}$. Multiply
   $5 \cdot 5$ to get a numerator of 25. The denominator remains 7.
   The result is $\frac{25}{7}$. Change this to the mixed number of $3\frac{4}{7}$
   by dividing 7 into 25. Your answer, as a mixed number,
   is $3\frac{4}{7}$.

4. Change $7\frac{1}{2}$ to an improper fraction: $2 \cdot 7 + 1 = 15$ is the
   numerator, and 2 remains as the denominator. The fraction
   becomes $\frac{15}{2}$. Change $2\frac{2}{9}$ to an improper fraction:
   $9 \cdot 2 + 2 = 20$ is the numerator, and 9 remains the denominator. The fraction becomes $\frac{20}{9}$. The problem now looks like
   $\frac{15}{2} \div \frac{20}{9}$. Invert $\frac{20}{9}$ to its reciprocal of $\frac{9}{20}$ and change the
   problem to multiplication, $\frac{15}{2} \times \frac{9}{20}$. Since 5 divides evenly
   into both 15 and 20, $15 \div 5 = 3$ and $20 \div 5 = 4$, use can-

celing to simplify: $\dfrac{\overset{3}{\cancel{15}}}{2} \times \dfrac{9}{\underset{4}{\cancel{20}}}$. Multiply the numerators,

$3 \cdot 9 = 27$, and the denominators, $2 \cdot 4 = 8$. The result is

$\dfrac{27}{8}$. Change this to a mixed number by dividing 8 into 27.

Your answer is $3\dfrac{3}{8}$.

5. Change $1\dfrac{1}{6}$ to an improper fraction, $6 \cdot 1 + 1 = 7$ is the

numerator, and 6 remains as the denominator. The fraction

becomes $\dfrac{7}{6}$. The problem now reads $\dfrac{7}{6} \div \dfrac{3}{4}$. Invert the sec-

ond fraction to its reciprocal of $\dfrac{4}{3}$ and multiply. The problem

now looks like $\dfrac{7}{6} \times \dfrac{4}{3}$. Since 2 is a common factor to both

4 and 6, $4 \div 2 = 2$ and $6 \div 2 = 3$, you can use canceling to

simplify, $\dfrac{7}{\underset{3}{\cancel{6}}} \times \dfrac{\overset{2}{\cancel{4}}}{3}$. Multiply the numerators, $7 \cdot 2 = 14$, and

the denominators, $3 \cdot 3 = 9$. The result is $\dfrac{14}{9}$. To change $\dfrac{14}{9}$

to a mixed number, divide 9 into 14. The answer is $1\dfrac{5}{9}$.

6. Change $2\dfrac{3}{5}$ to the improper fraction $\dfrac{13}{5}$: $5 \cdot 2 + 3 = 13$,

and 5 remains as the denominator. Change $3\dfrac{3}{10}$ to the im-

proper fraction $\dfrac{33}{10}$: $10 \cdot 3 + 3 = 33$, and 10 remains as the

denominator. The problem now looks like $\dfrac{13}{5} \div \dfrac{33}{10}$. Invert

$\dfrac{33}{10}$ to its reciprocal of $\dfrac{10}{33}$ and multiply, $\dfrac{13}{5} \times \dfrac{10}{33}$. Since 5 is

a common factor to both 5 and 10, you can use canceling to

simplify, $\dfrac{13}{\underset{1}{\cancel{5}}} \times \dfrac{\overset{2}{\cancel{10}}}{33}$. Multiply $13 \cdot 2$ in the numerator to equal

26, and $1 \cdot 33$ in the denominator to equal 33. Your answer

is $\dfrac{26}{33}$.

7. Invert $\dfrac{1}{3}$ to its reciprocal of $\dfrac{3}{1}$, or 3. Change the problem to

multiplication, $\dfrac{6}{5} \times \dfrac{3}{1}$. Multiply $6 \cdot 3$ in the numerator to

equal 18, and $5 \cdot 1$ in the denominator to equal 5. You now

have the improper fraction $\dfrac{18}{5}$. Change this improper

fraction to the mixed number $3\dfrac{3}{5}$.

8. Invert 2 to its reciprocal of $\frac{1}{2}$ and multiply. The problem now looks like $\frac{7}{8} \times \frac{1}{2}$. Multiply to get $\frac{7 \times 1}{8 \times 2} = \frac{7}{16}$.

9. Change $2\frac{1}{4}$ to the improper fraction $\frac{9}{4}$: $4 \cdot 2 + 1 = 9$, and 4 remains as the denominator. Change $1\frac{3}{8}$ to the improper fraction, $\frac{11}{8}$: $8 \cdot 1 + 3 = 11$, and 8 remains the denominator. The problem now looks like $\frac{9}{4} \div \frac{11}{8}$. Invert $\frac{11}{8}$ to its reciprocal of $\frac{8}{11}$ and change the problem to multiplication. The problem now looks like $\frac{9}{4} \times \frac{8}{11}$. Use canceling to divide 4 into both 4 and 8, $\frac{9}{{}_1 4} \times \frac{8^{\,2}}{11}$. Multiply the numerators to get 18, and the denominators to get 11. You now have the improper fraction $\frac{18}{11}$. Divide 11 into 18 to get a mixed number answer of $1\frac{7}{11}$.

10. Change $2\frac{4}{5}$ to an improper fraction $\frac{14}{5}$. Now invert $\frac{14}{5}$ to its reciprocal of $\frac{5}{14}$ and change the problem to multiplication, $\frac{7}{3} \times \frac{5}{14}$. Use canceling to divide 7 into 14 and into itself, $\frac{{}^1 7}{3} \times \frac{5}{14_{\,2}}$. Multiply the remaining numerators and denominators to get an answer of $\frac{5}{6}$.

CHAPTER FIVE

Awesome Addition

What did $\frac{1}{2}$ of a bagel say to the other $\frac{1}{2}$?

"Together we make a whole."

SAME SUPER DENOMINATOR

When you add 3 apples to 2 apples you get 5 apples. When you add 5 oranges to 3 oranges you get 8 oranges. It's the same with fractions with the same denominator. When the denominators are the same (like the same kind of fruit), keep the denominators (like the fruit *name*) and add the numerators (like the *number* of apples or oranges). Then simplify your answer by reducing or by changing to a mixed number in simplest form.

Proper fractions

For example, add the proper fractions $\frac{1}{7} + \frac{3}{7}$.

1. Keep the 7 as your denominator (like the *name* apples or oranges).

2. Add the numerators (like the *number* of apples or oranges).

3. Your answer is $\frac{4}{7}$, which is in simplest form.

Mixed number and proper fraction

For another example, add the mixed number and the fraction $2\frac{3}{5} + \frac{1}{5}$.

1. Keep the 5 as your denominator (same fruit).

2. Add the two numerators.

$$\frac{3}{5} + \frac{1}{5} = \frac{4}{5}$$

3. In this case, the whole number remains unchanged, and becomes part of your answer.

4. Your answer is $2\frac{4}{5}$.

Whole number and mixed number

Can you add a whole number to a mixed number?

Sure, for example, add $5 + 3\frac{1}{7}$.

1. Keep the fraction the same and add the two whole numbers.

$$5 + 3 = 8$$

2. Your answer is $8\frac{1}{7}$.

Can the whole number be the second number?

Absolutely, for example, add $3\frac{1}{7} + 5$.

1. Keep the fraction the same and add the two whole numbers.

$$3 + 5 = 8$$

2. Your answer is $8\frac{1}{7}$.

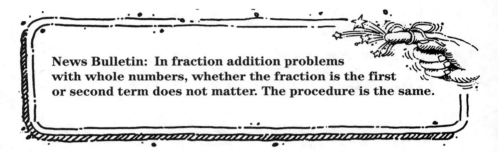

News Bulletin: In fraction addition problems with whole numbers, whether the fraction is the first or second term does not matter. The procedure is the same.

Two mixed numbers

What if you have two mixed numbers?

No problem at all. For example, add $3\frac{5}{8} + 2\frac{1}{8}$.

1. Keep the denominator the same (same fruit).

2. Add the two numerators.

$$\frac{5}{8} + \frac{1}{8} = \frac{6}{8}$$

3. Add the two whole numbers.

$$3 + 2 = 5$$

4. Your answer at this point is $5\frac{6}{8}$.

5. Reduce the fraction by dividing 6 and 8 by 2, $\frac{\cancel{6}^3}{\cancel{8}_4}$.

6. The answer is $5\frac{3}{4}$.

For another example, add $3\frac{2}{3} + 1\frac{1}{3}$.

1. Keep the denominator the same (same fruit again).

2. Add the two numerators.

$$\frac{2}{3} + \frac{1}{3} = \frac{3}{3}$$

3. Add the two whole numbers.

$$3 + 1 = 4$$

4. The answer now is $4\frac{3}{3}$.

5. Since $\frac{3}{3}$ equals 1, $4\frac{3}{3} = 4 + 1 = 5$.

What about $2\frac{3}{5} + \frac{4}{5}$?

1. Keep the 5 as your denominator (same fruit).

2. Add the two numerators.

$$\frac{3}{5} + \frac{4}{5} = \frac{7}{5}$$

3. You now have $2\frac{7}{5}$.

4. Since $\frac{7}{5}$ is an improper fraction, change it to the mixed number $1\frac{2}{5}$.

5. The problem now looks like $2 + 1\frac{2}{5}$.

6. Add the whole numbers and keep the fraction to get an answer of $3\frac{2}{5}$.

For yet another easy example, add $4\frac{2}{7} + 5\frac{6}{7}$.

1. Keep the denominator the same (same fruit once again).

2. Add the two numerators.

$$\frac{2}{7} + \frac{6}{7} = \frac{8}{7}$$

3. Add the two whole numbers.

$$4 + 5 = 9$$

4. Your answer at this point is $9\frac{8}{7}$.

5. Since $\frac{8}{7}$ equals $1\frac{1}{7}$, simplify by adding the whole numbers and bringing along the fraction.

$$9\frac{8}{7} = 9 + 1\frac{1}{7}$$

6. Your answer is $10\frac{1}{7}$.

SILLY:

$2\frac{8}{3}$ (a mixed number containing an improper fraction)

SUPERB:

$4\frac{2}{3}$ (From SILLY above, change the improper fraction $\frac{8}{3}$ to the mixed number $2\frac{2}{3}$. Then add $2 + 2\frac{2}{3}$ to get $4\frac{2}{3}$.)

DIFFERENT DELIGHTFUL
DENOMINATORS

When the denominators are delightfully different (like oranges
and apples), you must find a common denominator, the LCM
(like mixing the fruit together in a fruit salad).

Change each of the original denominators into the LCM by building equivalent fractions. Now, since the denominators are the
same,

• keep the new denominator,
• add the numerators and whole numbers, and
• simplify your answer.

Proper fractions

Is it easy to work with two proper fractions that have different denominators?

Yes it is. For example, add $\frac{5}{12} + \frac{1}{3}$.

1. Find the LCM for 12 and 3 because they are different denominators (different fruit). The LCM equals 12 because 12 is the lowest number that both 3 and 12 can divide into evenly.

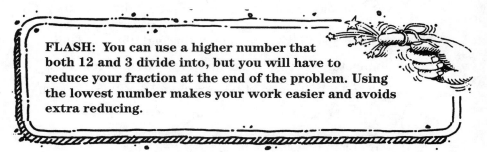

> **FLASH:** You can use a higher number that both 12 and 3 divide into, but you will have to reduce your fraction at the end of the problem. Using the lowest number makes your work easier and avoids extra reducing.

2. Rewrite each of the original fractions with the common denominator by building an equivalent fraction for each one, $\frac{5}{12} = \frac{5}{12}$ and $\frac{1}{3} = \frac{4}{12}$.

3. The problem looks like $\frac{5}{12} + \frac{4}{12}$.

4. Keep the denominator the same (now it's the same fruit).

5. Add the two numerators.

$$\frac{5}{12} + \frac{4}{12} = \frac{9}{12}$$

6. Your answer at this point is $\frac{9}{12}$.

7. Simplify your answer by dividing 9 and 12 by 3.

$$9 \div 3 = 3 \qquad \frac{\cancel{9}^{3}}{\cancel{12}_{4}}$$

$$12 \div 3 = 4$$

8. Your answer in lowest terms is $\frac{3}{4}$.

Mixed numbers

What about mixed numbers, the ones with a whole number and a fraction?

For example, add $3\frac{2}{7} + 1\frac{1}{6}$.

1. Calculate the LCM for 7 and 6. List the multiples of 7: 7, 14, 21, 28, 35, 42, 49, 56, 63 . . . . List the multiples of 6: 6, 12, 18, 24, 30, 36, 42 . . . . The LCM equals 42 because 42 is the lowest common multiple that both 7 and 6 divide into evenly.

2. Rewrite each of the original fractions with the common denominator by building an equivalent fraction for each one; $3\frac{2}{7} = 3\frac{12}{42}$ because $42 \div 7 = 6$ and $6 \times 2 = 12$, and $1\frac{1}{6} = 1\frac{7}{42}$ because $42 \div 6 = 7$ and $7 \times 1 = 7$.

3. The problem now looks like $3\frac{12}{42} + 1\frac{7}{42}$.

4. Keep the denominator the same.

5. Add the two numerators.

$$\frac{12}{42} + \frac{7}{42} = \frac{19}{42}$$

6. Add the two whole numbers.

$$3 + 1 = 4$$

7. Your answer is $4\frac{19}{42}$ which is in simplest form.

Three or more fractions

Food for thought: You may be wondering if three's a crowd here—not with fractions. Two or more fractions can be added. There is no maximum to the number of fractions that can be added. You just need to find an LCM for *all* your denominators.

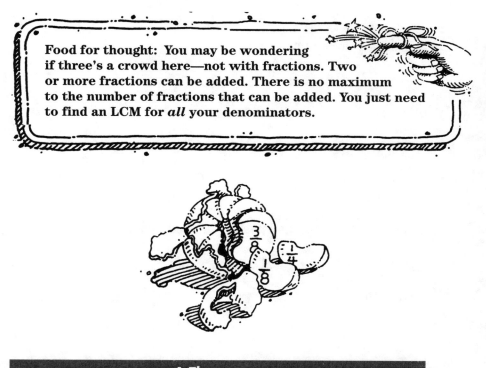

A Threesome

Add $1\frac{1}{2} + \frac{4}{5} + 2\frac{3}{4}$.

1. Find the LCM for all three denominators. List the multiples of 2: 2, 4, 6, 8, 10, 12, 14, 16, 18, 20, 22 . . . . List the multiples of 4: 4, 8, 12, 16, 20 . . . . List the multiples of 5: 5, 10, 15, 20, 25, 30 . . . . The LCM is 20 because 20 is the lowest common multiple that 2, 4, and 5 can divide into evenly.

2. Rewrite each of the original fractions with the common denominator by building an equivalent fraction for each one.

   $1\frac{1}{2} = 1\frac{10}{20}$ because $20 \div 2 = 10$ and $10 \times 1 = 10$.

   $\frac{4}{5} = \frac{16}{20}$ because $20 \div 5 = 4$ and $4 \times 4 = 16$.

   $2\frac{3}{4} = 2\frac{15}{20}$ because $20 \div 4 = 5$ and $5 \times 3 = 15$.

3. The problem now looks like $1\frac{10}{20} + \frac{16}{20} + 2\frac{15}{20}$.

4. Keep the denominator the same.

5. Add the three numerators.

$$\frac{10}{20} + \frac{16}{20} + \frac{15}{20} = \frac{41}{20}$$

6. Add the two whole numbers.

$$1 + 2 = 3$$

7. Your answer at this point is $3\frac{41}{20}$, but it needs some more work.

8. Change the fractional part to a mixed number.

$$41 \div 20 = 2\frac{1}{20}$$

9. Add the $2\frac{1}{20}$ to the 3.

$$2\frac{1}{20} + 3 = 5\frac{1}{20}$$

Your answer is $5\frac{1}{20}$.

If your answer is something like $5\frac{6}{6}$ or $5\frac{7}{6}$, change the fractional part to a whole number or a mixed number and simplify as for problems with the same denominator.

$$5\frac{6}{6} = 5 + 1 = 6 \qquad 5\frac{7}{6} = 5 + 1\frac{1}{6} = 6\frac{1}{6}$$

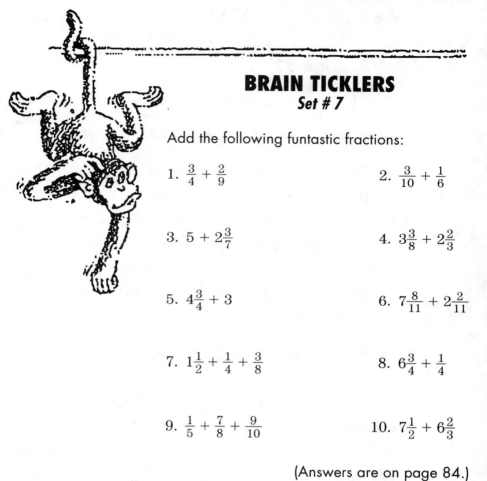

BRAIN TICKLERS
Set # 7

Add the following funtastic fractions:

1. $\frac{3}{4} + \frac{2}{9}$

2. $\frac{3}{10} + \frac{1}{6}$

3. $5 + 2\frac{3}{7}$

4. $3\frac{3}{8} + 2\frac{2}{3}$

5. $4\frac{3}{4} + 3$

6. $7\frac{8}{11} + 2\frac{2}{11}$

7. $1\frac{1}{2} + \frac{1}{4} + \frac{3}{8}$

8. $6\frac{3}{4} + \frac{1}{4}$

9. $\frac{1}{5} + \frac{7}{8} + \frac{9}{10}$

10. $7\frac{1}{2} + 6\frac{2}{3}$

(Answers are on page 84.)

BRAIN TICKLERS—THE ANSWERS

Set # 7, page 83

For all of these brain ticklers, find a common denominator that all of your denominators can divide into evenly. The best common denominator is the LCM, or least common multiple.

1. List the multiples of 4: 4, 8, 12, 16, 20, 24, 28, 32, 36 . . . .
   List the multiples of 9: 9, 18, 27, 36 . . . . The LCM is 36.
   Rewrite $\frac{3}{4}$ with a denominator of 36, $\frac{3}{4} = \frac{?}{36}$. Divide 4
   into 36 and multiply the result by 3 to fill in the ?,
   $36 \div 4 = 9$ and $9 \times 3 = 27$. So, $\frac{3}{4} = \frac{27}{36}$. Similarly, $\frac{2}{9} = \frac{8}{36}$.
   The problem now looks like $\frac{27}{36} + \frac{8}{36}$. Add the numerators.
   The answer is $\frac{35}{36}$.

2. List the multiples of 10: 10, 20, 30, 40 . . . . List the multiples of 6: 6, 12, 18, 24, 30 . . . . The LCM is 30. Rewrite $\frac{3}{10}$
   with the denominator of 30, $\frac{3}{10} = \frac{?}{30}$. Divide 10 into 30
   and multiply the result by 3 to fill in the ?, $30 \div 10 = 3$
   and $3 \times 3 = 9$. So, $\frac{3}{10} = \frac{9}{30}$. Similarly, $\frac{1}{6} = \frac{5}{30}$. The problem
   now looks like $\frac{9}{30} + \frac{5}{30}$. Add the numerators to get $\frac{14}{30}$. Since
   2 is a common factor to both 14 and 30, divide 2 into both
   the numerator and denominator, $\frac{\overset{7}{14}}{\underset{15}{30}} = \frac{7}{15}$. Your answer in
   simplest form is $\frac{7}{15}$.

3. In this problem, a common denominator is not necessary
   since 5 is a whole number. Add the two whole numbers,
   $5 + 2 = 7$, and bring along the fraction. Your answer is $7\frac{3}{7}$.

4. List the multiples of 8: 8, 16, 24 . . . . List the multiples of
   3: 3, 6, 9, 12, 15, 18, 21, 24 . . . . The LCM is 24. Rewrite $\frac{3}{8}$
   with a denominator of 24, $\frac{3}{8} = \frac{?}{24}$. Divide 8 into 24 and
   multiply the result by 3 to fill in the ?, $24 \div 8 = 3$ and
   $3 \times 3 = 9$. So, $\frac{3}{8} = \frac{9}{24}$. Similarly, $\frac{2}{3} = \frac{16}{24}$. The problem
   now looks like $3\frac{9}{24} + 2\frac{16}{24}$. Add the whole numbers,
   $3 + 2 = 5$, and your fractions $\frac{9}{24} + \frac{16}{24} = \frac{25}{24}$. You now
   have $5\frac{25}{24}$. Change the $\frac{25}{24}$ to a mixed number by dividing 24
   into 25 to get $1\frac{1}{24}$. You now have $5 + 1\frac{1}{24}$. Add the two
   whole numbers and bring along the fraction to get an an-
   swer of $6\frac{1}{24}$.

5. Here, a common denominator is unnecessary. Simply add
   the two whole numbers and bring along the fraction,
   $4\frac{3}{4} + 3 = 7\frac{3}{4}$.

6. Since the denominators are the same, you can keep this de-
   nominator and add the two numerators, $\frac{8}{11} + \frac{2}{11} = \frac{10}{11}$. Then
   add the two whole numbers, $7 + 2 = 9$. Your answer is $9\frac{10}{11}$.

7. Here you have three different denominators, so list the multiples for each one. The multiples of 2 are 2, 4, 6, 8 . . . , the
   multiples of 4 are 4, 8, 12 . . . , and the multiples of 8 are 8,
   16 . . . . The LCM for all three is 8. Rewrite $\frac{1}{2}$ with a denominator of 8, $\frac{1}{2} = \frac{?}{8}$. Divide 2 into 8 and multiply the result by
   1 to fill in the ?, $8 \div 2 = 4$ and $4 \times 1 = 4$. So, $\frac{1}{2} = \frac{4}{8}$. Similarly, $\frac{1}{4} = \frac{2}{8}$. The fraction $\frac{3}{8}$ stays the same because the denominator is already 8. The problem now looks like $1\frac{4}{8} + \frac{2}{8}$
   $+ \frac{3}{8}$. Since there is only one whole number, add the numerators of the fractions and bring along the whole number to
   get $1\frac{9}{8}$. Now change $\frac{9}{8}$ to a mixed number by dividing 8 into
   9 to get $1\frac{1}{8}$. You now have $1 + 1\frac{1}{8}$. Add the two whole numbers and bring along the fraction to get an answer of $2\frac{1}{8}$.

8. Since the denominators are the same, keep this denominator and add the two numerators, $\frac{3}{4} + \frac{1}{4} = \frac{4}{4}$. Since $\frac{4}{4} = 1$, add this 1 to the whole number 6. Your answer is 7.

9. List the multiples for each of the three denominators. The multiples of 5 are 5, 10, 15, 20, 25, 30, 35, 40 . . . , the multiples of 8 are 8, 16, 24, 32, 40 . . . , and the multiples of 10 are 10, 20, 30, 40 . . . . The LCM is 40. Rewrite $\frac{1}{5}$ with a denominator of 40, $\frac{1}{5} = \frac{?}{40}$. Divide 5 into 40 and multiply the result by 1 to fill in the ?, $40 \div 5 = 8$ and $8 \times 1 = 8$. So, $\frac{1}{5} = \frac{8}{40}$. Similarly, $\frac{7}{8} = \frac{35}{40}$, and $\frac{9}{10} = \frac{36}{40}$. The problem now reads $\frac{8}{40} + \frac{35}{40} + \frac{36}{40}$. Add the numerators to equal $\frac{79}{40}$. Change $\frac{79}{40}$ to a mixed number by dividing 40 into 79 to get an answer of $1\frac{39}{40}$.

10. List the multiples of 2: 2, 4, 6, 8 . . . . List the multiples of 3: 3, 6, 9, 12 . . . . Here, 6 is the LCM. Rewrite $\frac{1}{2}$ with a denominator of 6, $\frac{1}{2} = \frac{?}{6}$. Divide 2 into 6 and multiply the result by 1 to fill in the ?, $6 \div 2 = 3$ and $3 \times 1 = 3$. So, $\frac{1}{2} = \frac{3}{6}$. Similarly, $\frac{2}{3} = \frac{4}{6}$. The problem now reads $7\frac{3}{6} + 6\frac{4}{6}$. Add the whole numbers, $7 + 6 = 13$, and the fractions, $\frac{3}{6} + \frac{4}{6} = \frac{7}{6}$. You now have $13\frac{7}{6}$. Rewrite $\frac{7}{6}$ as the mixed number $1\frac{1}{6}$. The problem now reads $13 + 1\frac{1}{6}$. Add the two whole numbers and bring along the fraction to get an answer of $14\frac{1}{6}$.

CHAPTER SIX

Sparkling Subtraction

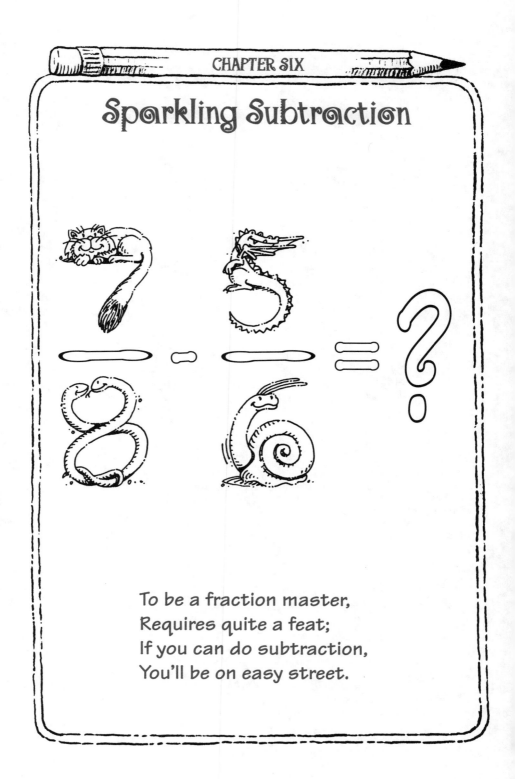

$$\frac{7}{8} - \frac{5}{6} = ?$$

To be a fraction master,
Requires quite a feat;
If you can do subtraction,
You'll be on easy street.

For most subtraction problems, the procedure is the same as that for adding fractions except that you *subtract* the numerators instead of *add* them.

SAME SUPER DENOMINATOR

For the **same denominator** (all apples *or* all oranges), keep the denominator and subtract the numerators and the whole numbers.

$$\frac{4}{7} - \frac{2}{7} = \frac{2}{7} \qquad\qquad 4\frac{4}{7} - 3\frac{1}{7} = 1\frac{3}{7}$$

DIFFERENT DELIGHTFUL DENOMINATORS

For **different denominators** (both apples *and* oranges), calculate a common denominator (the LCM—the fruit salad). Change each of the original denominators into the LCM by building equivalent fractions. Now, since the denominators are the same,

- keep the new denominator,
- subtract the numerators and whole numbers, and
- simplify your answer.

Spectacular Example

Subtract $\frac{6}{9} - \frac{1}{4}$.

1. Find the LCM of 9 and 4. The multiples of 9 are 9, 18, 27, 36 . . . , and the multiples of 4 are 4, 8, 12, 16, 20, 24, 28, 32, 36 . . . . The LCM equals 36 because 36 is the lowest number that both 9 and 4 can divide into evenly.

2. Rewrite each of the original fractions with the common denominator by building an equivalent fraction for each one; $\frac{6}{9} = \frac{24}{36}$ by dividing 36 by 9 and multiplying the result by 6.

$$36 \div 9 = 4 \qquad 4 \times 6 = 24$$

Also, $\frac{1}{4} = \frac{9}{36}$ by dividing 36 by 4 and multiplying the result by 1.

$$36 \div 4 = 9 \qquad 9 \times 1 = 9$$

3. The problem now looks like $\frac{24}{36} - \frac{9}{36}$.

4. Keep the denominator the same.

5. Subtract the two numerators.

$$24 - 9 = 15$$

6. Your answer at this point is $\frac{15}{36}$.

7. Reduce your answer (by dividing 15 and 36 by 3) to equal $\frac{5}{12}$, which is in simplest form.

● ●

Another Spectacular Example

Subtract $12\frac{5}{12} - 3\frac{1}{4}$.

1. Find the LCM of 12 and 4. The multiples of 12 are 12, 24, 36 . . . , and the multiples of 4 are 4, 8, 12, 16 . . . . The LCM equals 12 because 12 is the lowest number that both 12 and 4 can divide into evenly.

2. Rewrite each of the original fractions with the common denominator by building an equivalent fraction for each one: $12\frac{5}{12} = 12\frac{5}{12}$ because 12 is already the common denominator, and $3\frac{1}{4} = 3\frac{3}{12}$ because $12 \div 4 = 3$ and $3 \times 1 = 3$.

3. The problem now looks like $12\frac{5}{12} - 3\frac{3}{12}$.

4. Keep the denominator the same.

5. Subtract the two numerators.

$$\frac{5}{12} - \frac{3}{12} = \frac{2}{12}$$

6. Subtract the two whole numbers.

$$12 - 3 = 9$$

7. Your answer at this point is $9\frac{2}{12}$.

8. Reduce the fraction by dividing 2 and 12 by 2.

$$\frac{2 \div 2}{12 \div 2} = \frac{1}{6}$$

9. Your answer is $9\frac{1}{6}$.

• •

MIXED NUMBER AND WHOLE NUMBER (WITHOUT BORROWING)

Mixing bowl #1

The mixed number is first and the whole number is second. For example, subtract $8\frac{1}{2} - 3$.

1. When the fraction is in the first number, you can just bring it down to your answer.

2. Subtract the whole numbers.

$$8 - 3 = 5$$

3. Your answer is $5\frac{1}{2}$.

• •

MIXED NUMBER AND WHOLE NUMBER (WITH BORROWING)

Mixing bowl #2—Switch the ingredients

The whole number is first and the mixed number is second.

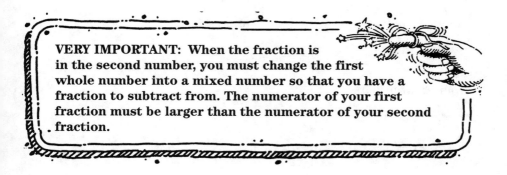

VERY IMPORTANT: When the fraction is in the second number, you must change the first whole number into a mixed number so that you have a fraction to subtract from. The numerator of your first fraction must be larger than the numerator of your second fraction.

For example, subtract $8 - 6\frac{1}{2}$.

1. Subtract 1 whole from the number 8, making it a 7.

2. Trade the 1 that you subtracted for a fraction with the same numerator and denominator as the denominator in the original problem, in this case, $\frac{2}{2}$.

3. Now, $8 = 7\frac{2}{2}$.

4. The problem now looks like $7\frac{2}{2} - 6\frac{1}{2}$.

5. Now proceed to subtract the whole numbers and the fractions, $7 - 6 = 1$ and $\frac{2}{2} - \frac{1}{2} = \frac{1}{2}$. Your answer is $1\frac{1}{2}$.

RIGHT MIX:

$6\frac{1}{3} - 4 = 2\frac{1}{3}$

MESSED UP MIX:

$6 - 4\frac{1}{3} = 2\frac{1}{3}$

RIGHT MIX:

$6 - 4\frac{1}{3} = 5\frac{3}{3} - 4\frac{1}{3} = 1\frac{2}{3}$

MIXED NUMBERS (WITH BORROWING)

What happens when you have mixed numbers where the second numerator is larger than the first numerator?

Let's try one where the denominators are the same.

For example (same fruit again), subtract $3\frac{1}{5} - 2\frac{4}{5}$.
You must trade or borrow here, too, but in a slightly different way from the previous problem.

Change the first mixed number into an equivalent form where the first fraction will be larger than the second fraction.

1. As before, subtract 1 whole from the first number, 3, making it a 2.

2. Trade the whole that you subtracted for a fraction with the same numerator and denominator as the denominator in the original problem, or $\frac{5}{5}$ in this case.

3. Now, $3 = 2\frac{5}{5}$.

4. Add this $\frac{5}{5}$ to the $\frac{1}{5}$ that is already there.

5. The first fraction now looks like $3\frac{1}{5} = 2\frac{5}{5} + \frac{1}{5} = 2\frac{6}{5}$.

6. The second fraction, $2\frac{4}{5}$, stays the same.

7. The problem now looks like $2\frac{6}{5} - 2\frac{4}{5}$.

8. Now subtract the whole number and the fractions.

$$2 - 2 = 0 \qquad \frac{6}{5} - \frac{4}{5} = \frac{2}{5}$$

9. Your answer is $\frac{2}{5}$.

Alternate Method

$3\frac{1}{5} - 2\frac{4}{5}$

1. Change both of the mixed numbers into improper fractions:
   $3\frac{1}{5} = \frac{16}{5}$ and $2\frac{4}{5} = \frac{14}{5}$.

2. Subtract $\frac{16}{5} - \frac{14}{5}$.

3. The answer is $\frac{2}{5}$.

SPARKLING SUBTRACTION

Now let's try one where the denominators are different fruit. First select a common denominator.

For example, subtract $8\frac{1}{5} - 4\frac{3}{10}$.

1. Change the denominators into a common one. Find the LCM. The multiples of 5 are 5, 10, 15 . . . , and the multiples of 10 are 10, 20 . . . . The LCM, the common denominator, equals 10 because 10 is the lowest number that both 5 and 10 can divide into evenly.

2. Rewrite each of the original fractions with the common denominator: $8\frac{1}{5} = 8\frac{2}{10}$, by dividing 5 into 10 and multiplying the result by 1, and $4\frac{3}{10} = 4\frac{3}{10}$ because 10 is already the common denominator.

3. Then you will notice that the second fraction is larger than the first, $8\frac{2}{10} - 4\frac{3}{10}$. So you must borrow or trade.

4. With $8\frac{2}{10}$, subtract 1 from the 8 making it a 7.

5. Trade the 1 that you subtracted for a fraction with the same numerator and denominator as the denominator in the original problem. This 1 is equal to $\frac{10}{10}$. Now $8 = 7\frac{10}{10}$.

6. Add this $\frac{10}{10}$ to the $\frac{2}{10}$ which is already there.

$$8\frac{2}{10} = 7\frac{10}{10} + \frac{2}{10} = 7\frac{12}{10}$$

7. The problem now looks like $7\frac{12}{10} - 4\frac{3}{10}$.

8. Subtract the numerators.

$$\frac{12}{10} - \frac{3}{10} = \frac{9}{10}$$

9. Keep the denominators the same.

10. Subtract the whole numbers.

$$7 - 4 = 3$$

11. The answer is $3\frac{9}{10}$.

Alternate Method

$8\frac{1}{5} - 4\frac{3}{10}$

1. Change both of the mixed numbers into improper fractions, $8\frac{1}{5} = \frac{41}{5}$ and $4\frac{3}{10} = \frac{43}{10}$.

2. The problem now looks like $\frac{41}{5} - \frac{43}{10}$.

3. Find an LCM for 5 and 10. The multiples of 5 are 5, 10, 15, 20 . . . , and the multiples of 10 are 10, 20, 30 . . . . The LCM, the common denominator, equals 10 because 10 is the lowest number that both 5 and 10 can divide into evenly.

4. Rewrite $\frac{41}{5}$ with the common denominator: $\frac{41}{5} = \frac{82}{10}$ by dividing 5 into 10 and multiplying the result by 41. $\frac{43}{10}$ already has 10 as the common denominator.

5. The problem now looks like $\frac{82}{10} - \frac{43}{10}$.

6. Subtract the numerators and keep the denominators the same, $\frac{82}{10} - \frac{43}{10} = \frac{39}{10}$.

7. Simplify $\frac{39}{10}$ by changing it to a mixed number of $3\frac{9}{10}$.

ADDITION-SUBTRACTION COMBO

Now, let's try a *SUPER DUPER* combination addition-subtraction problem.

For example, $\frac{7}{8} + \frac{1}{4} - \frac{1}{2}$

1. Change the denominators into the common one. Find an LCM. List the multiples of 8: 8, 16 . . . . List the multiples of 4: 4, 8, 12 . . . . List the multiples of 2: 2, 4, 6, 8 . . . . The LCM is 8 because 8 is the lowest number that 2, 4, and 8 can divide into evenly.

2. Rewrite each of the original fractions with the new common denominator.

   $\frac{7}{8} = \frac{7}{8}$ because 8 is already the common denominator.

   $\frac{1}{4} = \frac{2}{8}$ because $8 \div 4 = 2$ and $2 \times 1 = 2$.

   $\frac{1}{2} = \frac{4}{8}$ because $8 \div 2 = 4$ and $4 \times 1 = 4$.

3. The problem now looks like $\frac{7}{8} + \frac{2}{8} - \frac{4}{8}$.

4. Since the denominators are the same, add your two numerators $\frac{7}{8} + \frac{2}{8} = \frac{9}{8}$ and subtract the other numerator, $\frac{9}{8} - \frac{4}{8} = \frac{5}{8}$. The final answer is $\frac{5}{8}$.

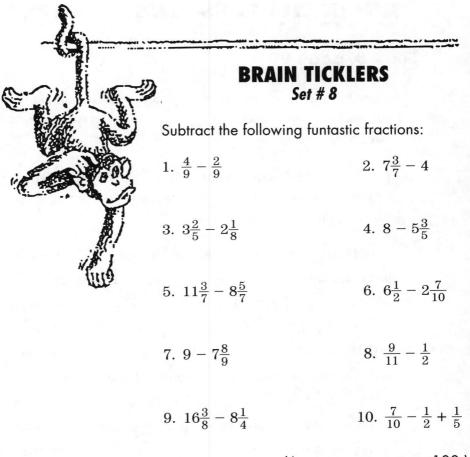

BRAIN TICKLERS
Set # 8

Subtract the following funtastic fractions:

1. $\frac{4}{9} - \frac{2}{9}$

2. $7\frac{3}{7} - 4$

3. $3\frac{2}{5} - 2\frac{1}{8}$

4. $8 - 5\frac{3}{5}$

5. $11\frac{3}{7} - 8\frac{5}{7}$

6. $6\frac{1}{2} - 2\frac{7}{10}$

7. $9 - 7\frac{8}{9}$

8. $\frac{9}{11} - \frac{1}{2}$

9. $16\frac{3}{8} - 8\frac{1}{4}$

10. $\frac{7}{10} - \frac{1}{2} + \frac{1}{5}$

(Answers are on page 100.)

BRAIN TICKLERS—THE ANSWERS

Set # 8, page 99

1. Since both fractions have the same denominator, simply subtract the two numerators, $\frac{4}{9} - \frac{2}{9} = \frac{4-2}{9}$ to get an answer of $\frac{2}{9}$.

2. Here, since the fraction is in the first part of the problem, you can subtract the two whole numbers and simply bring the fraction along to the answer, $7\frac{3}{7} - 4 = 3\frac{3}{7}$.

3. In order to subtract, you must find a common denominator for 5 and 8. List the multiples of 5: 5, 10, 15, 20, 25, 30, 35, 40, 45 . . . . List the multiples of 8: 8, 16, 24, 32, 40, 48 . . . . The LCM, the common denominator, is 40. Rewrite $\frac{2}{5}$ with the new denominator of 40, $\frac{2}{5} = \frac{?}{40}$. Divide 5 into 40 and multiply the result by 2. Because $40 \div 5 = 8$ and $8 \times 2 = 16$, $\frac{2}{5} = \frac{16}{40}$. Similarly, $\frac{1}{8} = \frac{5}{40}$ because $40 \div 8 = 5$ and $5 \times 1 = 5$. You now have $3\frac{16}{40} - 2\frac{5}{40}$. Since the denominators are now the same, and the first fraction is larger than the second, you just subtract the whole numbers and the fractions to get the answer of $1\frac{11}{40}$.

4. To complete the subtraction, you must change the 8 into a mixed number with 5 as the denominator of the fraction in order to match the other denominator of 5 already in the problem. To do this, subtract 1 from the number 8 making it a 7. This 1 that you subtracted or traded is equal to $\frac{5}{5}$. So, $8 = 7\frac{5}{5}$. The problem now looks like $7\frac{5}{5} - 5\frac{3}{5}$. Now, subtract the whole numbers and fractions to get the answer of $2\frac{2}{5}$.

5. Since the denominators are the same, you do not have to find a new common denominator. However, you cannot subtract the larger fraction of $\frac{5}{7}$ from the smaller fraction of $\frac{3}{7}$ in this form. You must change $11\frac{3}{7}$ to an equivalent mixed number so that it contains a fraction larger than $\frac{5}{7}$. To do this, subtract 1 whole from the 11, making it a 10. The 1 whole that you borrowed is equal to $\frac{7}{7}$ in this problem because of the 7 that is the denominator in the original problem. Add this $\frac{7}{7}$ to the $\frac{3}{7}$ that is already there, $11\frac{3}{7} = 10\frac{7}{7} + \frac{3}{7} = 10\frac{10}{7}$. Now the problem looks like $10\frac{10}{7} - 8\frac{5}{7}$. Subtract the whole numbers and the fractions. The answer is $2\frac{5}{7}$.

6. Find a common denominator for 2 and 10. The multiples of 2 are 2, 4, 6, 8, 10 . . . , and the multiples of 10 are 10, 20, 30 . . . . The LCM, which is the lowest common denominator, is 10. Rewrite $\frac{1}{2}$ with the new common denominator of 10, $\frac{1}{2} = \frac{?}{10}$. Because $10 \div 2$ and $5 \times 1 = 5$, the ? is equal to 5. So, $\frac{1}{2} = \frac{5}{10}$. The fraction $\frac{7}{10}$ is already written with the common denominator. The problem now looks like $6\frac{5}{10} - 2\frac{7}{10}$. Now you must rewrite $6\frac{5}{10}$ as an equivalent mixed number so that its fraction will be larger than $\frac{7}{10}$. Subtract 1 whole from the 6, making it a 5. The 1 whole that you subtracted can be traded for $\frac{10}{10}$, as $1 = \frac{10}{10}$. Add $\frac{10}{10}$ to the $\frac{5}{10}$ that is already there, $6\frac{5}{10} = 5\frac{10}{10} + \frac{5}{10} = 5\frac{15}{10}$. Now the problem looks like $5\frac{15}{10} - 2\frac{7}{10}$. Subtract the two whole numbers and the fractions to get an answer of $3\frac{8}{10}$. The fraction $\frac{8}{10}$ can be reduced to $\frac{4}{5}$ by dividing both the numerator and denominator by the common factor of 2, $\frac{\overset{4}{\cancel{8}}}{\underset{5}{\cancel{10}}}$. Your simplified answer is $3\frac{4}{5}$.

7. To complete the subtraction, you must change the 9 into a mixed number with 9 as the denominator of the fraction in order to match the other denominator of 9 already in the problem. To do this, subtract 1 whole from the 9, making it an 8. This 1 whole that you subtracted or traded is equal to $\frac{9}{9}$. So, $9 = 8\frac{9}{9}$, and $7\frac{8}{9}$ stays the same. Now subtract the two whole numbers and the two fractions, $8\frac{9}{9} - 7\frac{8}{9} = 1\frac{1}{9}$. Your answer is $1\frac{1}{9}$.

8. In this problem, you must find a common denominator for 11 and 2. The multiples of 11 are 11, 22, 33 . . . , and the multiples of 2 are 2, 4, 6, 8, 10, 12, 14, 16, 18, 20, 22 . . . . The LCM, the common denominator, is 22. Rewrite $\frac{9}{11}$ with the new common denominator of 22, $\frac{9}{11} = \frac{?}{22}$. Divide 11 into 22 and multiply the result by 9. Because $22 \div 11 = 2$ and $2 \times 9 = 18$, $\frac{9}{11} = \frac{18}{22}$. Similarly, $\frac{1}{2} = \frac{11}{22}$ because $22 \div 2 = 11$ and $11 \times 1 = 11$. The problem now looks like $\frac{18}{22} - \frac{11}{22}$. Subtract the two numerators to get the answer of $\frac{7}{22}$.

9. Here you need a common denominator for 8 and 4. The multiples of 8 are 8, 16, 24 . . . , and the multiples of 4 are 4, 8, 12 . . . . The LCM, the common denominator, is 8. Rewrite $\frac{1}{4}$ with the common denominator of 8, $\frac{1}{4} = \frac{?}{8}$. Divide 4 into 8 and multiply the result by 1. Because $8 \div 4 = 2$ and $2 \times 1 = 2$, $\frac{1}{4} = \frac{2}{8}$. The fraction $\frac{3}{8}$ stays the same because it is already written with the common denominator. The problem now looks like $16\frac{3}{8} - 8\frac{2}{8}$. Subtract the two whole numbers and the two fractions to get the answer of $8\frac{1}{8}$.

10. Although this problem has both addition and subtraction, one common denominator will cover all three fractions. The multiples of 10 are 10, 20 . . . , the multiples of 2 are 2, 4, 6, 8, 10 . . . , and the multiples of 5 are 5, 10, 15 . . . . The LCM, the common denominator, is 10. Rewrite $\frac{1}{2}$ with the new common denominator of 10, $\frac{1}{2} = \frac{?}{10}$. Divide 2 into 10 and multiply the result by 1 to get $\frac{1}{2} = \frac{5}{10}$. Similarly, $\frac{1}{5} = \frac{2}{10}$. The fraction $\frac{7}{10}$ stays the same because it already is written with the common denominator of 10. The problem now looks like $\frac{7}{10} - \frac{5}{10} + \frac{2}{10}$. Subtract the first two numerators, $7 - 5 = 2$, and then add the third numerator, $2 + 2 = 4$. Keep the denominator of 10. Your answer so far is $\frac{4}{10}$. Reduce $\frac{4}{10}$ by dividing the common factor of 2 into both the numerator and denominator, $\frac{\overset{2}{\cancel{4}}}{\underset{5}{\cancel{10}}}$. Your final answer in simplest form is $\frac{2}{5}$.

CHAPTER SEVEN

The Dreaded and Disliked Word Problems

Roses are red,
Violets are blue,
If you can't figure out a
whole word problem,
A fraction at a time will do.

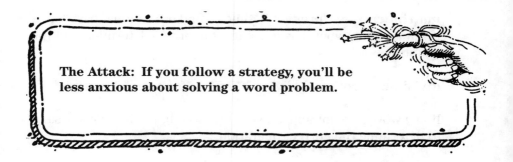

The Attack: If you follow a strategy, you'll be less anxious about solving a word problem.

STRATEGY FOR WORD PROBLEMS—PREVIEW, VIEW, AND REVIEW

Preview

Skim the problem once just to get the general idea. Don't write anything down.

View

Read the problem carefully looking for important information. Look for key words or phrases, such as those below, that may appear in your problem.

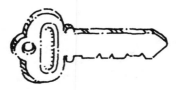

addition: "sum," "total," "altogether," "combine"

subtraction: "difference," "how much less," "how much more," "amount of decrease," "amount of increase," "how much is left"

multiplication: "product," "times," find a fractional part "of" a given amount, when given a single quantity—determine multiple quantities

division: "quotient," "divide," "for 1," "for each," "separate"

If key words do not appear, some reasoning might be necessary to determine the correct operation.

Plan your work. Write down the necessary steps to solve the problem. Estimate your answer. Solve it.

Review

Check your work. See if your answer makes sense and fits reasonably with the rest of the problem.

Wordy Examples

Addition

Juanita is a waitress at a diner. She works $5\frac{1}{2}$ hours on Friday, $2\frac{1}{4}$ hours on Saturday, and 6 hours on Sunday. How many hours *altogether* does she work?

1. "Altogether" is a key word for addition. Add $5\frac{1}{2}$, $2\frac{1}{4}$, and 6. Find a common denominator for 2 and 4. Use the LCM, 4, because it's the lowest number that both 2 and 4 can divide into evenly.

2. Rewrite the fractions with the common denominator: $5\frac{1}{2} = 5\frac{2}{4}$ because $4 \div 2 = 2$ and $2 \times 1 = 2$, $2\frac{1}{4} = 2\frac{1}{4}$ because 4 is already the common denominator.

3. Add the whole numbers and the fractions.

   $5 + 2 + 6 = 13$ $\qquad\qquad$ $\frac{2}{4} + \frac{1}{4} = \frac{3}{4}$

4. The answer is $13\frac{3}{4}$ hours.

Subtraction

Carlos can drive his new sports car from Raywood to Pigeon Hollow in $6\frac{2}{3}$ hours. If he has already driven $2\frac{3}{4}$ hours, *how much more* driving time *is left*?

1. The key words "how much more is left?" imply subtraction. Set up the problem with the larger whole number first.

$$6\frac{2}{3} - 2\frac{3}{4}$$

2. Find a common denominator for 3 and 4. Use their LCM, 12, for the common denominator.

3. $6\frac{2}{3} = 6\frac{8}{12}$ because $12 \div 3 = 4$ and $4 \times 2 = 8$; $2\frac{3}{4} = 2\frac{9}{12}$ because $12 \div 4 = 3$ and $3 \times 3 = 9$.

4. At this point, you have $6\frac{8}{12} - 2\frac{9}{12}$. In order to complete the subtraction, change the problem so that the first fraction is larger. You must subtract 1 whole from 6 to equal 5. The 1 whole that you subtracted or borrowed here is equal to $\frac{12}{12}$. The first fraction already has $\frac{8}{12}$ so add $\frac{12}{12} + \frac{8}{12} = \frac{20}{12}$. Now $6\frac{8}{12} = 5\frac{20}{12}$.

5. The problem now looks like $5\frac{20}{12} - 2\frac{9}{12}$.

6. Subtract the whole numbers and the fractions.

$$5\frac{20}{12} - 2\frac{9}{12} = 3\frac{11}{12}$$

7. The answer is $3\frac{11}{12}$ hours.

Multiplication

Jodie bought 100 shares of stock in the Lucky Circle cereal company. The price was \$$16\frac{3}{4}$ *a share.* How much did Jodie pay for the stock (not counting commission or any other costs)?

1. Since the price for *1* share is given and you want to know the price for *100* shares, use multiplication. Multiply $100 \cdot 6\frac{3}{4}$.

2. Rewrite $16\frac{3}{4}$ as an improper fraction by multiplying $16 \cdot 4 = 64$ and adding 3, $64 + 3 = 67$. 67 is your numerator and 4 remains as your denominator, $16\frac{3}{4} = \frac{67}{4}$.

3. Now you can multiply $100 \cdot \frac{67}{4}$. Use canceling to divide 4 into 100 and 4 into itself, $\overset{25}{100} \cdot \frac{67}{4}_1$. Multiply $25 \cdot 67$. Your answer is \$1675.

Division

Sima needs $\frac{1}{2}$ pound of ground turkey meat for each turkey burger she's making for the snack bar at the school fair. She has 16 pounds of turkey meat. Into how many turkey burgers can Sima *divide* her meat?

1. The word "divide" implies division. Start with the larger number, $16 \div \frac{1}{2}$.

2. Invert the second number and change to a multiplication problem, $16 \cdot \frac{2}{1}$.

3. Multiply $16 \cdot 2$ and simplify.

4. In this case, $16 \cdot 2 = 32$ turkey burgers.

• •

BRAIN TICKLERS
Set # 9

Solve these funtastic fraction word problems:
(Remember—Look for key words and use
strategies shown.)

1. Douglas poured $3\frac{1}{2}$ cups of flour into his
   mixing bowl. If his chocolate chip cookie
   recipe calls for 5 cups of flour, *how much
   more* flour does he need?

2. Lisette has $12\frac{3}{4}$ feet of wood and wants to *divide* it into 3
   shelves. How long will each shelf be?

3. Joan baby-sits 4 hours on Monday, $3\frac{1}{2}$ hours on Tuesday, $5\frac{1}{3}$
   hours on Wednesday, 5 hours on Thursday, and $4\frac{1}{4}$ hours on
   Friday. How many hours does Joan baby-sit *altogether* for
   the 5 days?

4. Jesse can mow the lawn in 6 hours. It takes his brother Tim $\frac{3}{4}$ *of that time*. How long does it take Tim to mow the lawn?

5. Erica worked on her math homework for $\frac{7}{8}$ of an hour and her English homework for $\frac{3}{4}$ of an hour. For how many hours did Erica work on math and English *combined*?

6. Janice and Judy are twins. At birth, Judy weighed 7 pounds and Janice weighed $6\frac{5}{8}$ pounds. What was the *difference* in their weight?

7. Jordan's jeep travels $22\frac{1}{2}$ miles on *one* gallon of gasoline. How far can it travel on $4\frac{1}{2}$ gallons of gasoline?

8. Michael's mom bought a turkey that weighed $12\frac{3}{4}$ pounds for the holidays. After the skin and the breast bone were *removed*, the turkey weighed $10\frac{1}{2}$ pounds.

a) What was the weight of the skin and bone?

b) Into how many $\frac{1}{4}$ pound servings can Michael's mom *divide* the turkey?

9. After finishing his homework, Ethan sat down in front of his big screen TV to watch the football game. At the time, team A had $\frac{2}{7}$ *of the number of points* that team B had. If team B had 21 points, how many did team A have?

10. Rosalie was recording her favorite songs from the radio onto cassette tapes. She completed 2 tapes, $\frac{2}{3}$ of another, and $\frac{3}{5}$ of a fourth tape. *In total*, how many tapes contain music?

11. The Santiago family purchased stock in Smooth Airlines one year ago. The stock was priced at $5\frac{1}{8}$ points. Today each share of stock is worth $7\frac{1}{2}$ points. How many points did each share *increase*?

12. Mrs. Cohen's math class was working on a stock market project. If they bought one share of stock in a toy company for $\$12\frac{3}{8}$, one share of stock in a sporting goods company for $\$20\frac{1}{4}$, and one share of stock in a record company for $\$5\frac{7}{8}$, how much did they spend for the three shares of stock *altogether*?

(Answers are on page 116.)

BRAIN TICKLERS—THE ANSWERS

Set # 9, page 115

Look for key words or phrases to make your work easier.

1. The phrase "how much more" gives a hint of subtraction. To find how much more flour, subtract the amount used from the total amount needed. The problem becomes $5 - 3\frac{1}{2}$. You must change the 5 into a mixed number so that you can subtract fractions. Here, subtract 1 from 5 to equal 4. The 1 whole that you borrowed is equal to $\frac{2}{2}$ because 2 is the denominator already in the problem. Since $5 = 4\frac{2}{2}$ and $3\frac{1}{2} = 3\frac{1}{2}$, the problem now looks like $4\frac{2}{2} - 3\frac{1}{2}$. Now subtract the whole numbers and the fractions, $4\frac{2}{2} - 3\frac{1}{2} = 1\frac{1}{2}$. The answer is $1\frac{1}{2}$ more cups of flour.

2. The word "divide" lets you know this is a division problem. Set your problem up with the larger whole number first, $12\frac{3}{4} \div 3$. Invert the 3 to its reciprocal and set up a multiplication problem, $12\frac{3}{4} \times \frac{1}{3}$. Change $12\frac{3}{4}$ to an improper fraction by multiplying 4 times 12 and adding 3, $4 \times 12 = 48$ and $48 + 3 = 51$. Keep your denominator of 4, so $12\frac{3}{4} = \frac{51}{4}$. The problem now looks like $\frac{51}{4} \times \frac{1}{3}$. Use canceling to divide 3 into 51 and 3 into 3, $\frac{\overset{17}{51}}{4} \times \frac{1}{\underset{1}{3}}$. Multiply the remaining numbers to get $\frac{17}{4}$. Divide 4 into 17 to get a mixed number answer of $4\frac{1}{4}$. Each shelf will be $4\frac{1}{4}$ feet long.

3. The word "altogether" implies addition. Add $4 + 3\frac{1}{2} + 5\frac{1}{3} + 5 + 4\frac{1}{4}$. You can add the whole numbers, $4 + 3 + 5 + 5 + 4 = 21$, and find a common denominator for the three fractions $\frac{1}{2}, \frac{1}{3}$, and $\frac{1}{4}$. The multiples of 2 are 2, 4, 6, 8, 10, 12 . . . , the multiples of 3 are 3, 6, 9, 12 . . . , and the multiples of 4 are 4, 8, 12 . . . . The LCM is 12. Rewrite $\frac{1}{2}$ with the common denominator of 12, $\frac{1}{2} = \frac{?}{12}$. Divide 2 into 12 and multiply the result by 1. The new result, 6, fills in the ?, so $\frac{1}{2} = \frac{6}{12}$. Similarly, $\frac{1}{3} = \frac{4}{12}$ and $\frac{1}{4} = \frac{3}{12}$. Your fractions now look like $\frac{6}{12} + \frac{4}{12} + \frac{3}{12}$. Add them to equal $\frac{13}{12}$. Put together the whole number answer and the fraction answer to equal $21\frac{13}{12}$. Change $\frac{13}{12}$ to $1\frac{1}{12}$ by dividing 12 into 13 with a fractional remainder. You now have $21 + 1\frac{1}{12}$. Add the whole numbers and bring along the fraction to equal $22\frac{1}{12}$ hours of babysitting for Joan.

4. The key words "of that time" imply multiplication, since you are finding a fractional part of a given amount. Multiply $6 \times \frac{3}{4}$. Use canceling to divide 2 into both 6 and 4, $\overset{3}{\cancel{6}} \times \frac{3}{\underset{2}{\cancel{4}}}$. Multiply the remaining numerators and denominators to get $\frac{9}{2}$. Change $\frac{9}{2}$ into a mixed number by dividing 2 into 9. The answer is $4\frac{1}{2}$. It takes Tim $4\frac{1}{2}$ hours to mow the lawn.

5. The word "combined" gives a hint of an addition problem. Add $\frac{7}{8}$ of an hour to $\frac{3}{4}$ of an hour, $\frac{7}{8} + \frac{3}{4}$. Find a common denominator for 8 and 4. The multiples of 8 are 8, 16 . . . , and the multiples of 4 are 4, 8, 12 . . . . . The LCM, the common denominator, is 8. Change $\frac{3}{4}$ into $\frac{6}{8}$ by dividing 4 into 8 and multiplying the result by 3 to get 6, $\frac{3}{4} = \frac{6}{8}$. The fraction $\frac{7}{8}$ stays the same because it is already written with the common denominator. You now have $\frac{7}{8} + \frac{6}{8}$. Add the numerators to get $\frac{13}{8}$. Change $\frac{13}{8}$ to a mixed number by dividing 8 into 13 to get 1 with a fractional remainder of $\frac{5}{8}$. Erica worked on her math and English homework for a combined $1\frac{5}{8}$ hours.

6. The word "difference" is the key word for the answer to a subtraction problem. The difference in Janice and Judy's weight is the answer to $7 - 6\frac{5}{8}$. Change 7 into a mixed number with a fractional denominator of 8 so that you have two fractional parts to subtract. To do this, subtract 1 whole from the 7, making it a 6. In this problem, the 1 that you borrowed is equal to $\frac{8}{8}$. Since $7 = 6\frac{8}{8}$ and $6\frac{5}{8} = 6\frac{5}{8}$, the problem now looks like $6\frac{8}{8} - 6\frac{5}{8}$. Subtract the whole numbers and the fractions.

$$6\frac{8}{8} - 6\frac{5}{8} = \frac{3}{8}$$

The answer is $\frac{3}{8}$ of a pound difference in weight.

7. Since the number of miles for one gallon is given, multiply to find the number of miles for $4\frac{1}{2}$ gallons. The problem looks like $22\frac{1}{2} \times 4\frac{1}{2}$. Change $22\frac{1}{2}$ to an improper fraction by multiplying 2×22 and adding 1, $2 \times 22 = 44$ and $44 + 1 = 45$. Keep your denominator of 2, so $22\frac{1}{2} = \frac{45}{2}$. For $4\frac{1}{2}$, multiply 2×4 and add 1, $2 \times 4 = 8$ and $8 + 1 = 9$. Keep your denominator of 2, so $4\frac{1}{2} = \frac{9}{2}$. The problem now looks like $\frac{45}{2} \times \frac{9}{2}$. There are no common factors between the numerators and the denominators, $45 \times 9 = 405$ and $2 \times 2 = 4$. The answer as an improper fraction is $\frac{405}{4}$. Divide 4 into 405 to get the mixed number $101\frac{1}{4}$. Jordan's jeep can travel $101\frac{1}{4}$ miles on $4\frac{1}{2}$ gallons of gasoline.

8. a) From the word "removed," you get a hint of subtraction. Subtract the weight of the turkey after the trim from the weight of the turkey before the trim. This gives $12\frac{3}{4} - 10\frac{1}{2}$. The LCM of 4 and 2, your new common denominator, is 4. The mixed number $12\frac{3}{4}$ already has the new denominator of 4. For $10\frac{1}{2}$, divide 2 into 4 and multiply the resulting answer of 2 by the 1 in the numerator to get 2, $10\frac{1}{2} = 10\frac{2}{4}$. Now, subtract $12\frac{3}{4} - 10\frac{2}{4}$ to get $2\frac{1}{4}$. The total weight of the skin and breast bone was $2\frac{1}{4}$ pounds.

8. b) The key word "divide" refers to division. Take the current weight of the turkey, $10\frac{1}{2}$ pounds, and divide this weight by $\frac{1}{4}$, which is the weight of each serving, $10\frac{1}{2} \div \frac{1}{4}$. Change $10\frac{1}{2}$ into an improper fraction by multiplying 2×10 and adding 1, $2 \times 10 + 1 = 21$. This 21 becomes your numerator and 2 remains as your denominator to give $\frac{21}{2}$. Invert $\frac{1}{4}$ to its reciprocal of 4 and change the problem to multiplication, $\frac{21}{2} \times 4$. Divide 2 into both 2 and 4, $\frac{21}{1\cancel{2}} \times \cancel{4}^{2}$. Multiply the remaining numbers to get $\frac{42}{1}$. Your answer is 42. Michael's mom can divide the turkey into 42 servings.

9. The phrase "$\frac{2}{7}$ of the number of points that team B had" implies multiplication, since you are finding a fractional part of a given amount. The given amount is team B's score of 21. Multiply $\frac{2}{7} \times 21$. Use canceling to divide 7 into both 21 and itself, $\frac{2}{1\cancel{7}} \times \cancel{21}^{3}$. Multiply the remaining numerators to get 6. Team A had 6 points.

10. The word "total" refers to addition. Add $2 + \frac{2}{3} + \frac{3}{5}$ by finding a common denominator for 3 and 5. The multiples of 3 are 3, 6, 9, 12, 15 . . . , and the multiples of 5 are 5, 10, 15. . . . The LCM, the common denominator, is 15. Rewrite $\frac{2}{3}$ by dividing 3 into 15 and multiplying the result by 2 to get $\frac{2}{3} = \frac{10}{15}$. Similarly, $\frac{3}{5} = \frac{9}{15}$. You now have $2 + \frac{10}{15} + \frac{9}{15}$. Add the two fractions to get $2 + \frac{19}{15}$. Change $\frac{19}{15}$ into a mixed number by dividing 15 into 19 to get 1 with a fractional remainder of $\frac{4}{15}$. So, $\frac{19}{15} = 1\frac{4}{15}$. You now have $2 + 1\frac{4}{15}$. Add the whole numbers and bring the fractional part along to get $3\frac{4}{15}$. Rosalie used $3\frac{4}{15}$ tapes.

11. To find the "amount of increase" subtract $5\frac{1}{8}$ from $7\frac{1}{2}$: $7\frac{1}{2} - 5\frac{1}{8}$. Find a common denominator by listing the multiples of 2: 2, 4, 6, 8, 10 . . . , and the multiples of 8: 8, 16, 24 . . . . The LCM and the common denominator are 8. Rewrite $7\frac{1}{2}$ with the new denominator of 8, $7\frac{1}{2} = 7\frac{?}{8}$. Divide 2 into 8 and multiply this answer of 4 by the 1 in the numerator to get 4, $7\frac{1}{2} = 7\frac{4}{8}$. $5\frac{1}{8}$ already has the common denominator, so subtract $7\frac{4}{8} - 5\frac{1}{8}$ to get $2\frac{3}{8}$. Each share of stock increased by $2\frac{3}{8}$ points.

12. The word "altogether" indicates addition. Add $12\frac{3}{8}$, $20\frac{1}{4}$, and $5\frac{7}{8}$. In order to find a common denominator, list multiples of 4: 4, 8, 12, 16 . . . , and multiples of 8: 16, 24, 32 . . . . The common denominator and LCM are 8. Rewrite $20\frac{1}{4}$ with the new denominator of 8, $20\frac{1}{4} = 20\frac{?}{8}$. Divide 4 into 8 and multiply the answer of 2 by 1, $20\frac{1}{4} = 20\frac{2}{8}$. The other fractions of $12\frac{3}{8}$ and $5\frac{7}{8}$ already have the common denominator. Now you can add $12\frac{3}{8} + 20\frac{2}{8} + 5\frac{7}{8}$. Add the whole numbers $12 + 20 + 5 = 37$, and the fractions, $\frac{3}{8} + \frac{2}{8} + \frac{7}{8} = \frac{12}{8}$. Change $\frac{12}{8}$ into $1\frac{4}{8} = 1\frac{1}{2}$ by dividing 8 into 12 and reducing to simplest form. The answer so far is $37 + 1\frac{1}{2}$. Add to get a final answer of $\$38\frac{1}{2}$.

CHAPTER EIGHT

Refreshing Ratios, Remarkable Rates, and Popular Proportions

1 to 2, 3 to 4
Rates and proportions
Are fun to explore.

REFRESHING RATIOS

A ratio of two numbers is a comparison of the numbers using the same units. A ratio can be expressed as a fraction, with a colon, or with the word "to." Ratios should be expressed in simplest form, where the numbers in the numerator and denominator have no common factors.

AWESOME:

A ratio of 1 to 2 = 1 : 2 = $\frac{1}{2}$

AWFUL:

A ratio of 4 to 3 = 4 : 3 = $\frac{4}{3}$.

Exquisite Example

What is the ratio of $3 to $5 (dollars is the same unit for both parts)? Since the units are the same, the label of the units can be eliminated in your answer. The answer can be expressed as 3 to 5, 3 : 5, or $\frac{3}{5}$.

Another Exquisite Example

What is the ratio of 4 coins to 12 coins? The answer can be expressed as 4 to 12, 4 : 12, or $\frac{4}{12}$. Since the numbers 4 and 12 both have common factors of 2 and 4, you can use either to simplify your fraction.

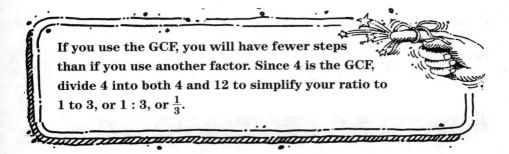

If you use the GCF, you will have fewer steps than if you use another factor. Since 4 is the GCF, divide 4 into both 4 and 12 to simplify your ratio to 1 to 3, or 1 : 3, or $\frac{1}{3}$.

Still Another Exquisite Example

What is the ratio of 6 cups to 2 cups? The answer can be expressed as 6 to 2, 6 : 2, or $\frac{6}{2}$. Since 6 and 2 have a common GCF of 2, you can divide 2 into both 6 and 2 to simplify your ratio to 3 to 1, 3 : 1, or $\frac{3}{1}$.

And Yet Another Exquisite Example

What is the ratio of 8 hours to 1 day? Change "one day" into 24 hours so that the ratio can be compared in the same units. The answer can then be expressed as 8 to 24, 8 : 24, or $\frac{8}{24}$. Since 8 and 24 have a greatest common factor of 8, simplify your ratio to 1 to 3, 1 : 3, or $\frac{1}{3}$.

• •

NOTE: Hereafter in this book we will express ratios in the fraction form only.

Ratios can also be used in word problems. The word or words that come before the word "to" form the numerator of your ratio, and the word or words after the word "to" form the denominator of the ratio.

Beautiful Example

Tae rode his bike 10 miles on Saturday and 12 miles on Sunday. What is the ratio of his miles on Saturday to his miles on Sunday?

Set up a ratio: $\dfrac{\text{miles on Saturday}}{\text{miles on Sunday}}$

The miles on Saturday are 10 and the miles on Sunday are 12, so your fractional ratio becomes $\frac{10}{12}$. Since 2 is a common factor to both 10 and 12, you can use canceling $\frac{\cancel{10}^{5}}{\cancel{12}_{6}}$, to simplify your fraction to $\frac{5}{6}$. Therefore, the ratio of Tae's miles on Saturday to Tae's miles on Sunday is $\frac{5}{6}$.

● ●

Satisfying Example

Loribeth's winning softball team scored 14 runs in the championship game. The losing team scored 8 runs.

a) What is the ratio of the winning score to the losing score?

Set up a ratio: $\dfrac{\text{winning score}}{\text{losing score}}$

The winning score is equal to 14 runs and the losing score is equal to 8 runs, so your fractional ratio becomes $\frac{14}{8}$. Since 2 is a common factor to both 14 and 8, you can use canceling, $\frac{\overset{7}{14}}{\underset{4}{8}}$, to simplify your fraction to $\frac{7}{4}$. Therefore, the ratio of the winning score to the losing score is $\frac{7}{4}$.

b) What is the ratio of the number of winning runs scored *to* the *total* number of runs scored?

Set up a ratio: $\dfrac{\text{winning runs scored}}{\text{total runs scored}}$

The winning runs scored are equal to 14 and the total runs scored can be obtained by adding winning runs plus losing runs. Since the number of winning runs is 14 and the number of losing runs is 8, the total number of runs is equal to $14 + 8 = 22$. The fractional ratio is $\frac{14}{22}$. Since 2 is a common factor to both 14 and 22, you can use canceling, $\frac{14}{22}\frac{7}{11}$, to simplify your fraction to $\frac{7}{11}$. Therefore, the ratio of the winning score to the total score is $\frac{7}{11}$.

. . . .

Pleasurable Example

600 people attended the school play on Friday night. 350 people attended on Saturday night. What is the ratio of attendance on Friday *to* the attendance on Saturday?

Set up a ratio: $\dfrac{\text{Friday's attendance}}{\text{Saturday's attendance}}$

Friday's attendance is equal to 600 and Saturday's attendance is equal to 350, so the fractional ratio is $\frac{600}{350}$. Since 50 is a common factor to both 600 and 350, you can use canceling, $\frac{600}{350}\frac{12}{7}$, to reduce your fraction to $\frac{12}{7}$. Therefore, the ratio of Friday's attendance to Saturday's attendance is $\frac{12}{7}$.

. . . .

BRAIN TICKLERS
Set # 10

Write each comparison as a ratio in simplest form using a fraction:

1. 30 minutes to 1 hour

2. 8 days to 20 days

3. 75 cents to 50 cents

4. 10 feet to 10 feet

5. $640 to $820

Set up and solve each word problem as a ratio in simplest form:

6. Ms. Foschi bought a television in an appliance store for $560. Mr. Green bought his television in a department store for $620. Find the ratio of the price of the appliance store television *to* the price of the department store television.

7. Steve's dad bought a computer 3 years ago for $2,000. To-day the computer is worth $800 due to depreciation. Find the ratio of the computer's depreciation (decrease in value) *to* its original cost.

8. The home economics class used 5 cups of sugar, 3 cups of flour, and 2 cups of water to stir into its cupcake batter.

   a) What is the ratio of sugar *to* water?
   b) What is the ratio of flour *to* the total ingredients listed above?

9. The price of a candy bar increased from $.35 to $.65 in a year. What is the ratio of the increase in price *to* the original price?

10. Micky can type 40 words per minute while his sister can type only 28 words per minute. What is the ratio of Micky's typing speed *to* his sister's typing speed?

(Answers are on page 145.)

REMARKABLE RATES

A **rate** is a comparison of two quantities using *different* units. It is written as a fraction.

RIGHT:

$$\frac{8 \text{ pounds of meat}}{5 \text{ people}} \qquad \frac{12 \text{ pounds of grass seed}}{5 \text{ trees}}$$

WRONG:

$$\frac{8 \text{ girls}}{7 \text{ girls}}$$ (This is a *ratio* with the *same* units.)

Rates should be expressed in simplest form, where the numbers in the numerator and denominator have no common factors.

A special kind of rate, known as a **unit rate**, is one in which the denominator is 1.

CORRECT:

$$\frac{15 \text{ miles}}{1 \text{ hour}}$$

CRUMMY:

$$\frac{15 \text{ miles}}{4 \text{ hours}}$$

To find a unit rate, divide your denominator into your numerator.

GREAT:

$$\frac{\$15.60}{3 \text{ hour}} = \frac{\$5.20}{1 \text{ hour}}$$

GRAND:

$$\frac{6 \text{ cupcakes}}{3 \text{ people}} = \frac{2 \text{ cupcakes}}{1 \text{ person}}$$

GRACEFUL:

$$\frac{78 \text{ gallons of gas}}{4 \text{ miles}} = \frac{19.5 \text{ gallons of gas}}{1 \text{ mile}}$$

Rates can also be used in word problems. The words and numbers that come before prepositional words such as "to," "for," "on," and "in" form the numerator of your rate written as a fraction, and the word or words after these prepositional words form the denominator of your rate written as a fraction. Unit rate word problems often contain words "per" or "each."

There are two ways to put your rate in simplest form.

1. Divide the GCF into both the numerator and denominator. Simplify the fraction.

2. Divide the denominator into the numerator to get a unit rate, a rate with 1 unit in the denominator.

. In many problems the rate in simplest form is the unit rate.

Here are some examples of writing rates in simplest form:

DANDY:

2400 words on 120 pages

Set up a rate: $\dfrac{2400 \text{ words}}{120 \text{ pages}}$

Divide 120 into 2400. $120\overline{)2400}$.

$$\begin{array}{r} 20 \\ 120\overline{)2400} \end{array}$$

The new rate is $\dfrac{20 \text{ words}}{1 \text{ page}}$

The rate of 20 words on 1 page is the unit rate, which is in simplest form.

DELIGHTFUL:

28 children in 8 cars

Set up a rate: $\dfrac{28 \text{ children}}{8 \text{ cars}}$

The GCF between 28 and 8 is 4, so divide 4 into both the numerator and denominator.

$$\frac{28 \text{ children}}{8 \text{ cars}} = \frac{7 \text{ children}}{2 \text{ cars}}$$

The rate in simplest form is 7 children in 2 cars. Also, if you divide 2 into 7,

you get an answer of $3\frac{1}{2}$. So, $3\frac{1}{2}$ children in 1 car is your *unit* rate. Both ways to express your rate in simplest form are correct, but because $\frac{1}{2}$ child doesn't make sense, it is better to leave your rate as 7 children in 2 cars in this problem.

WORTHY:

Jené works at a gas station after school and on weekends. She worked 24 hours last week and earned \$146.40 for the week. How much did she make *per* hour?

The word "per" refers to a rate with 1 unit in the denominator, or a unit rate. Divide the number of hours into the total salary for the week to find the salary per hour,

$$
\begin{array}{r}
6.10 \\
24\overline{)146.40}
\end{array}
$$

Jené's salary per hour is \$6.10.

JOLLY:

The total cost of making 1150 pairs of jeans in the factory is \$10,450. If 150 pairs are discarded due to irregularities, what is the cost *per* pair of jeans for those that are not discarded?

The word "per" refers to a unit rate. First subtract the number of discards from the total number of jeans, $1150 - 150 = 1000$. Now, divide the number of jeans that are not discarded into the total price of making all the jeans, $10,450 \div 1000 = 10.45$. The cost per pair of jeans that are not discarded is \$10.45.

BRAIN TICKLERS
Set # 11

Write as a rate or unit rate in simplest form:

1. $56 for 4 shirts

2. 75 cents for 5 pieces of gum

3. 168 miles in 4 hours

4. 30 books on 8 shelves

5. 250 trees planted on 5 acres

Set up and solve each word problem as a rate in simplest form:

6. The Cityview Movie Theater charges $19.80 for 4 adult matinee movie tickets. How much does each ticket cost?

7. A toy company sold 550 mini-footballs for $4174.50. What is the price per football?

8. A new plane travels 2100 miles in $2\frac{1}{2}$ hours. Find the plane's speed in miles per hour.

9. The total cost of producing 2000 baseball cards was $220.80. Of these cards, 160 had to be discarded due to imperfections. What is the cost per card of those baseball cards that were not discarded?

10. The students in Mrs. Jamison's English class want to read 77 books in $3\frac{1}{2}$ weeks. How many books per week should they read?

(Answers are on page 147.)

POPULAR PROPORTIONS

The equality of two ratios or rates forms a *proportion*.

TRUE:

$$\frac{1}{2} = \frac{2}{4}$$

The cross products of a proportion are equal, $\frac{1}{2} = \frac{2}{4}$, $1 \cdot 4 = 2 \cdot 2$.

CORRECT:

$$\frac{8 \; pounds}{10 \; pounds} = \frac{4 \; pounds}{5 \; pounds}$$

EQUIVALENT:

$$\frac{6 \; girls}{7 \; boys} = \frac{12 \; girls}{14 \; boys}$$

FALSE:

$\frac{1}{8} = \frac{1}{2}$ because $1 \cdot 8$ does not equal $1 \cdot 2$

Sometimes one of the numbers in a proportion is missing. We name this missing piece by a letter called a **variable**. You can solve the proportion for the missing piece or variable by following these steps:

1. Multiply the two cross products.

2. Set them equal to each other.

3. Divide the number on the same side as the variable into the number on the other side of the equation.

4. The result of step 3 is the missing piece to your proportion.

5. Check the cross products to make sure they are equal.

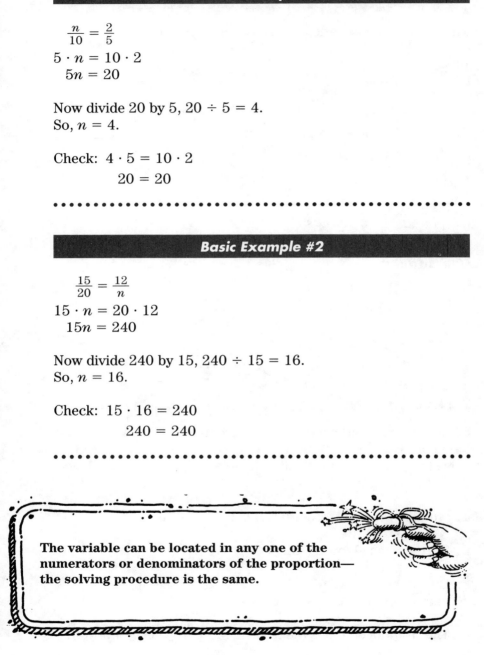

Basic Example #1

$$\frac{n}{10} = \frac{2}{5}$$
$$5 \cdot n = 10 \cdot 2$$
$$5n = 20$$

Now divide 20 by 5, $20 \div 5 = 4$.
So, $n = 4$.

Check: $4 \cdot 5 = 10 \cdot 2$
$$20 = 20$$

• •

Basic Example #2

$$\frac{15}{20} = \frac{12}{n}$$
$$15 \cdot n = 20 \cdot 12$$
$$15n = 240$$

Now divide 240 by 15, $240 \div 15 = 16$.
So, $n = 16$.

Check: $15 \cdot 16 = 240$
$$240 = 240$$

• •

The variable can be located in any one of the numerators or denominators of the proportion— the solving procedure is the same.

A Little Different Example

$$\frac{4}{n} = \frac{3}{7}$$
$$3 \cdot n = 4 \cdot 7$$
$$3n = 28$$

Now divide 28 by 3, $28 \div 3 = 9$, remainder 1.

$$
\begin{array}{r}
9 \\
3\overline{)28} \\
\underline{27} \\
1
\end{array}
$$

The remainder can be expressed as a fraction of $\frac{1}{3}$ (the remainder over the divisor).

So, $n = 9\frac{1}{3}$

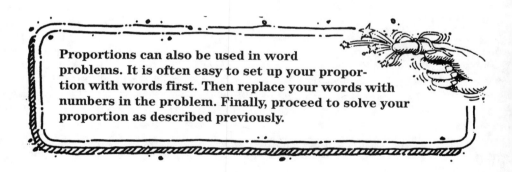

Proportions can also be used in word problems. It is often easy to set up your proportion with words first. Then replace your words with numbers in the problem. Finally, proceed to solve your proportion as described previously.

A Taxing Example

The sales tax on a $100 portable stereo purchase is $6. At this same rate, what is the tax on a $150 purchase? Set up your proportions as $\frac{tax}{purchase} = \frac{tax}{purchase}$. Replace your tax and purchase price as follows: $\frac{\$6}{\$100} = \frac{tax}{\$150}$ ($6 tax and $100 purchase price go together and the unknown tax and $150 go together. We'll call the unknown tax t). Multiply the cross products to get $6 \cdot 150 = 100 \cdot t$ or $900 = 100 \cdot t$. Divide 100 into 900 to get an answer of $9 tax. Therefore, you can conclude that a tax of $6 on a $100 purchase is equal to a tax of $9 on a $150 purchase.

A Delicious Example

A 16-ounce box of pasta contains 210 calories. How many calories are in a 2-ounce serving of pasta? Set up your proportion as $\frac{ounces\ of\ pasta}{calories} = \frac{ounces\ of\ pasta}{calories}$. Replace ounces of pasta and calories from the problem as follows: $\frac{16}{210} = \frac{2}{c}$ (16 ounces and 210 calories go together, and 2 ounces of pasta and the unknown calories go together; we'll call the unknown calories c). Multiply the cross products to get $16 \cdot c = 2 \cdot 210$, or $16c = 420$. Divide 16 into 420,

$$
\begin{array}{r}
26 \\
16\overline{)420} \\
\underline{32} \\
100 \\
\underline{96} \\
4
\end{array}
$$

to get an answer of $26\frac{4}{16}$. Reduce your fraction to get an answer of $26\frac{1}{4}$. Therefore, you can conclude that a 16-ounce box of pasta with 210 calories is equivalent to a 2-ounce serving of pasta with $26\frac{1}{4}$ calories.

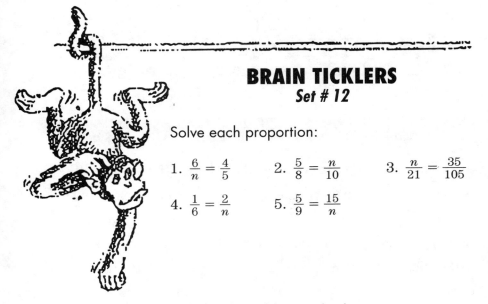

BRAIN TICKLERS
Set # 12

Solve each proportion:

1. $\frac{6}{n} = \frac{4}{5}$ 2. $\frac{5}{8} = \frac{n}{10}$ 3. $\frac{n}{21} = \frac{35}{105}$

4. $\frac{1}{6} = \frac{2}{n}$ 5. $\frac{5}{9} = \frac{15}{n}$

Set up a proportion for each problem and solve:

6. Torie gets paid $110 for every 20 hours she works at the ice cream store. At this rate, what is her total salary for 16 hours?

7. The junior baseball team won 21 games out of 30 that were played. At this rate, how many games would they have won if they played 40 games?

8. Tyeesha's new television cost $299. She paid a sales tax of $17.94. At this rate, what would be the sales tax on a $350 television?

9. On a map scale, $\frac{3}{4}$ of an inch is equal to 120 actual miles. At the same rate, how many actual miles would a $1\frac{1}{4}$-inch scale on a map represent?

10. It takes a school bus driver 75 minutes to drive a 20-mile route. At the same rate, how many minutes would it take to drive a 12-mile route?

(Answers are on page 151.)

BRAIN TICKLERS—THE ANSWERS

Set # 10, page 131

1. Since ratios must be in the same units, change 1 hour to 60 minutes. The ratio can be expressed as $\frac{30}{60}$ in fractional form. Since 30 is the GCF of 30 and 60, use canceling to divide 30 into both the numerator and denominator, $\frac{30\,^1}{60\,_2}$ to get a ratio of $\frac{1}{2}$ in simplest form.

2. The ratio can be expressed as $\frac{8}{20}$ in fractional form. Since 4 is the GCF of 8 and 20, you can use canceling to divide 4 into both the numerator and denominator, $\frac{8\,^2}{20\,_5}$, to get a ratio of $\frac{2}{5}$ in simplest form.

3. The ratio can be expressed as $\frac{75}{50}$ in fractional form. Since 25 is the GCF of 75 and 50, use canceling to divide 25 into both your numerator and denominator, $\frac{75\,^3}{50\,_2}$, to get a ratio of $\frac{3}{2}$ in simplest form.

4. The ratio can be expressed as $\frac{10}{10}$ in fractional form. Since the numerator and denominator are the same, your ratio becomes $\frac{1}{1}$, or 1 in simplest form.

5. The ratio can be expressed as $\frac{640}{820}$ in fractional form. Since both 640 and 820 end in zero, they are both divisible by 10. Dividing 10 into both the numerator and denominator produces $\frac{64}{82}$. Now, since 64 and 82 are both even numbers, use canceling to divide by 2, $\frac{64\,^{32}}{82\,_{41}}$ to get a ratio of $\frac{32}{41}$ in simplest form. If you had divided by the GCF of 20, instead of by the factor of 10, your work would have had fewer steps, but your result would be the same answer of $\frac{32}{41}$.

6. The price of the appliance store television becomes the numerator (clue: these are the words *before* the word "to") and the price of the department store television becomes the denominator (clue: these are the words *after* the word "to"). The ratio looks like $\frac{560}{620}$. Use canceling to divide 20 into both 560 and 620, $\frac{\cancel{560}\,28}{\cancel{620}\,31}$. Your ratio in simplest form is $\frac{28}{31}$.

7. To find the computer's depreciation, you must subtract the computer's current value from its original cost, $2000 - $800 = 1200. So, $1200 is the depreciation. Now set up a ratio with the depreciation in the numerator (clue: these are the words *before* the word "to") and the original cost in the denominator (clue: these are the words *after* the word "to"). The ratio looks like $\frac{1200}{2000}$. Now reduce the ratio to its simplest form by dividing by the GCF of 400, $\frac{\cancel{1200}\,3}{\cancel{2000}\,5}$. Your answer is $\frac{3}{5}$.

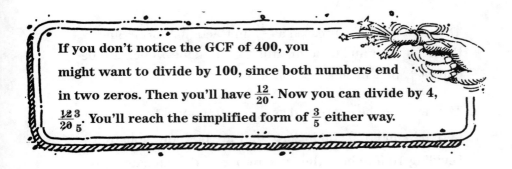

If you don't notice the GCF of 400, you might want to divide by 100, since both numbers end in two zeros. Then you'll have $\frac{12}{20}$. Now you can divide by 4, $\frac{\cancel{12}\,3}{\cancel{20}\,5}$. You'll reach the simplified form of $\frac{3}{5}$ either way.

8. a) The ratio would be represented as $\frac{sugar}{water}$. The number representing sugar is 5. This number becomes the numerator. The number representing water is 2. This number becomes the denominator. Your ratio is $\frac{5}{2}$. It is in simplest form.

b) The number representing flour becomes the numerator of 3, and the number representing total ingredients, $5 + 3 + 2 = 10$, becomes the denominator. Your ratio is $\frac{3}{10}$. It is in simplest form.

9. In this problem, $.35 is the original price of the candy and $.65 is the new price. To find the increase in price, subtract 0.35 from 0.65. The increase is 0.30. The ratio of the increase in price to the original price is $\frac{0.30}{0.35}$. Since 5 is the GCF of 30 and 35, divide 5 into both the numerator and the denominator and drop the decimal point representing cents, since the units are the same $\frac{30}{35}\,\frac{6}{7}$. The ratio in simplest form is $\frac{6}{7}$.

10. Micky's typing speed is in the numerator of the ratio and his sister's typing speed is in the denominator. The ratio looks like $\frac{40}{28}$. Since the GCF is 4, $\frac{40}{28}\,\frac{10}{7}$, and the fraction reduces to $\frac{10}{7}$ in simplest form.

Set # 11, page 137

1. Since the word "for" separates the numerator from your denominator, set up your problem as a rate.

$$\frac{\$56}{4 \text{ shirts}}$$

You can look for a GCF for 56 and 4, which is 4, $\frac{56}{4}\,\frac{14}{1}$. Instead of this, you can also divide 4 into 56 to get 14. The unit rate is $14 for 1 shirt.

2. Since the word "for" separates the numerator from the denominator, set up the problem as a rate.

$$\frac{75 \text{ cents}}{5 \text{ pieces of gum}}$$

You can write the rate in simplest form by dividing the GCF of 5 into both the numerator and denominator, $\frac{\overset{15}{\cancel{75}}}{\underset{1}{\cancel{5}}}$. Instead of this, you can also divide 5 into 75 to get a unit rate of 15 cents for 1 piece of gum.

3. Since the word "in" separates the numerator from the denominator, set up the problem as a rate.

$$\frac{168 \text{ miles}}{4 \text{ hours}}$$

You can simplify the rate by dividing the GCF of 4 into both your numerator and denominator, $\frac{\overset{42}{\cancel{168}}}{\underset{1}{\cancel{4}}}$. You can also get a unit rate by dividing 4 into 168. The answer is 42 miles in 1 hour in simplest form.

4. Since the word "on" separates the numerator from the denominator, set up the problem as a rate.

$$\frac{30 \text{ books}}{8 \text{ shelves}}$$

Divide the GCF of 2 into both your numerator and denominator, $\frac{\overset{15}{\cancel{30}}}{\underset{4}{\cancel{8}}}$. This produces a rate in simplest form of 15 books on 4 shelves. To get a unit rate, or the number of books on 1 shelf, divide 4 into 15 to get an answer of $3\frac{3}{4}$. The unit rate is $3\frac{3}{4}$ books on 1 shelf. Since $\frac{3}{4}$ of a book doesn't make sense, it would be better to keep the rate as 15 books on 4 shelves.

5. Since the word "on" separates the numerator from the denominator, set up the problem as a rate.

$$\frac{250 \text{ trees}}{5 \text{ acres}}$$

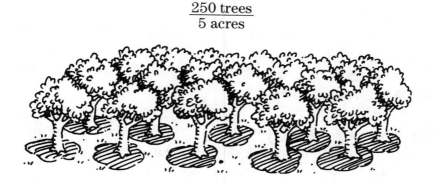

Since your GCF is 5, you can divide 5 into both your numerator and denominator, $\frac{250}{5}\frac{50}{1}$. You can also get a unit rate by dividing 5 into 250. The rate in simplest form is 50 trees planted on 1 acre.

6. The word "each" implies a unit rate. Divide 4 into $19.80,

$$\begin{array}{r} 4.95 \\ 4\overline{)19.80} \end{array}$$

to get a price of $4.95 for each ticket.

7. The word "per" implies a unit rate. Divide 550 into $4174.50,

$$\begin{array}{r} 7.59 \\ 550\overline{)4174.50} \end{array}$$

to get a unit rate of $7.59. The price per football is $7.59.

8. Divide 2100 by $2\frac{1}{2}$ to get a unit rate. To do this, write the division with $2\frac{1}{2}$ as an improper fraction, $2100 \div \frac{5}{2}$. Invert $\frac{5}{2}$ to its reciprocal of $\frac{2}{5}$ and turn the problem into multiplication, $2100 \times \frac{2}{5}$. Since 5 divides evenly into both 5 and 2100, use canceling to simplify the problem to $420 \times \frac{2}{1}$. Multiply the numerators, 420×2, to equal 840. The plane's speed is 840 miles per hour.

9. The word "per" indicates a unit rate. First, subtract the number of imperfect cards from the total number of baseball cards, $2000 - 160 = 1840$. Now, divide the number of cards that were not discarded into the total price of producing the cards,

$$1840 \overline{)220.80}^{\,0.12}$$

The cost per card of the baseball cards that were not discarded is $.12.

10. The word "per" implies a unit rate. Divide 77 by $3\frac{1}{2}$ to get a unit rate. To do this, write the division with $3\frac{1}{2}$ as an improper fraction, $77 \div \frac{7}{2}$. Invert $\frac{7}{2}$ to its reciprocal of $\frac{2}{7}$ and turn the problem into multiplication, $77 \times \frac{2}{7}$. Use canceling to divide 7 into both 7 and 77, $^{11}\cancel{77} \times \frac{2}{\cancel{7}_1}$. The problem simplifies to $11 \times \frac{2}{1}$. Multiply the numbers remaining to get 22. The class should read 22 books per week to meet its goal.

Set # 12, page 144

1. Multiply $4 \cdot n$ and set this equal to $6 \cdot 5$, $4n = 30$. Divide 4 into 30,

$$\begin{array}{r} 7 \\ 4\overline{)30} \\ \underline{28} \\ 2 \end{array}$$

to get an answer of $7\frac{2}{4}$. Reduce this to $7\frac{1}{2}$. Therefore, $n = 7\frac{1}{2}$.

2. Multiply $8 \cdot n$ and set this equal to $5 \cdot 10$, $8n = 50$. Divide 8 into 50,

$$\begin{array}{r} 6 \\ 8\overline{)50} \\ \underline{48} \\ 2 \end{array}$$

to get an answer of $6\frac{2}{8}$. Reduce this to $6\frac{1}{4}$. Therefore, $n = 6\frac{1}{4}$.

3. Multiply $105 \cdot n$ and set this equal to $21 \cdot 35$, $105n = 735$. Divide 105 into 735 to get an answer of 7. Therefore, $n = 7$.

4. Multiply $n \cdot 1$ and set this equal to $2 \cdot 6$, $1n = 12$. Therefore, $n = 12$.

5. Multiply $5 \cdot n$ and set this equal to $15 \cdot 9$, $5n = 135$. Divide 5 into 135,

$$\begin{array}{r} 27 \\ 5\overline{)135} \\ \underline{10} \\ 35 \\ \underline{35} \end{array}$$

to get an answer of 27. Therefore, $n = 27$.

6. Set up your proportion as $\frac{salary}{number\ of\ hours} = \frac{salary}{number\ of\ hours}$. Replace salary and number of hours from the problem as follows: $\frac{110}{20} = \frac{s}{16}$ ($110 and 20 hours go together and 16 hours and the unknown salary go together; we'll call the unknown salary s). Multiply the cross products to get $20 \cdot s = 110 \cdot 16$, or $20s = 1760$. Divide 20 into 1760,

$$
\begin{array}{r}
88 \\
20\overline{)1760} \\
\underline{160} \\
160 \\
\underline{160}
\end{array}
$$

to get an answer of 88. Therefore, Torie earns $88 for 16 hours.

7. Set up your proportion as $\frac{games\ won}{games\ played} = \frac{games\ won}{games\ played}$. Replace games won and games played from the problem as follows: $\frac{21}{30} = \frac{w}{40}$ (21 games won out of 30 games played go together and 40 games played and the unknown number of games won go together; we'll call the unknown number of games won w). Multiply the cross products to get $30 \cdot w = 21 \cdot 40$, or $30w = 840$. Divide 30 into 840,

$$
\begin{array}{r}
28 \\
30\overline{)840} \\
\underline{60} \\
240 \\
\underline{240}
\end{array}
$$

to get an answer of 28. Therefore, you can conclude that winning 21 games out of 30 is equivalent to winning 28 games out of 40.

8. Set up your proportion as $\dfrac{tax}{purchase\ price} = \dfrac{tax}{purchase\ price}$.

   Replace the tax and purchase price from the problem as follows: $\dfrac{\$17.94}{299} = \dfrac{t}{350}$ (The television for $299 goes with the sales tax of $17.94 and the television for $350 goes with the unknown tax t). Multiply the cross products to get $299 \cdot t = 17.94 \cdot 350$, or $299t = 6279$. Divide 299 into 6279,

$$
\begin{array}{r}
21 \\
299\overline{)6279} \\
\underline{598} \\
299 \\
\underline{299}
\end{array}
$$

   to get an answer of 21. Therefore, you can conclude that a sales tax of $21 on a $350 television purchase is equivalent to a sales tax of $17.94 on a $299 purchase.

9. Set up your proportion as $\dfrac{scale}{actual\ miles} = \dfrac{scale}{actual\ miles}$.
   Replace the scale and actual miles from the problem as follows:

$$
\dfrac{\frac{3}{4}}{120} = \dfrac{1\frac{1}{4}}{m}
$$

   ($\frac{3}{4}$ of an inch scale represents 120 miles and $1\frac{1}{4}$ inch represents the unknown number of miles m). Multiply the cross products to get $\frac{3}{4} \cdot m = 120 \cdot 1\frac{1}{4}$. Change $1\frac{1}{4}$ to an improper fraction to get $\frac{3}{4}m = 120 \cdot \frac{5}{4}$. Use canceling to divide 4 into both 4 and 120. Multiplying the remaining numbers gives $\frac{3}{4}m = 150$. Next, divide 150 by $\frac{3}{4}$,

$150 \div \frac{3}{4}$. To do this, invert $\frac{3}{4}$ to its reciprocal of $\frac{4}{3}$ and multiply $150 \cdot \frac{4}{3}$. Use canceling to divide 3 into 150 and 3 into itself, $\overset{50}{\cancel{150}} \cdot \frac{4}{\underset{1}{\cancel{3}}}$. Multiply the remaining numbers to get $m = 200$. Therefore, you can conclude that a distance of $1\frac{1}{4}$ inches on the map is equal to 200 actual miles.

10. Set up your proportion as $\frac{minutes}{miles} = \frac{minutes}{miles}$. Replace the minutes and miles from the problem as follows: $\frac{75}{20} = \frac{m}{12}$ (75 minutes goes with 20 miles and the unknown number of minutes m goes with 12 miles). Multiply the cross products to get $20 \cdot m = 75 \cdot 12$, or $20m = 900$. Divide 20 into 900,

$$
\begin{array}{r}
45 \\
20\overline{)900} \\
\underline{80} \\
100 \\
\underline{100}
\end{array}
$$

to get an answer of 45. Therefore, the bus driver would drive 12 miles in 45 minutes. This is equivalent to driving 20 miles in 75 minutes.

Exceptional Exponents

What did one exponent say to the other?

"Boy, we little numbers are power-ful!"

STRATEGY AND SAMPLES

An **exponent** is a little raised number to the right of another number, called the **base**. For example, in 2^3, 2 is the base and 3 is the exponent. The exponent tells how many times to multiply the base by itself. The 3 tells you how many times to multiply the 2 by itself. Therefore $2^3 = 2 \times 2 \times 2$, which equals 8. When whole numbers greater than 1 are raised to positive exponents greater than 1, the answer is a bigger number than the original base. When 1 is raised to any positive exponent, the answer is still 1.

Caution—Major Mistake Territory!

Do not say 2^3 equals 2 times 3 which equals 6.

You can use exponents with fractions. For example, in $\left(\frac{1}{2}\right)^2$, $\frac{1}{2}$ is the base and 2 is the exponent.

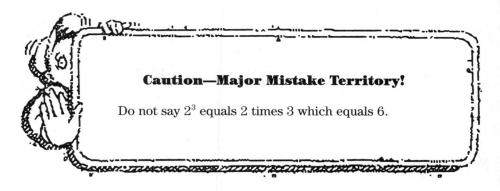

Parentheses are used here to separate the fraction from the exponent. The 2 tells you how many times to multiply the $\frac{1}{2}$ by itself. Therefore, $\left(\frac{1}{2}\right)^2$ looks like $\frac{1}{2} \cdot \frac{1}{2}$, which equals $\frac{1}{4}$. When proper fractions are raised to positive exponents greater than 1, the answer is a smaller number than the original base. When improper fractions or mixed numbers are raised to positive exponents greater than 1, the answer is a larger number than the original base.

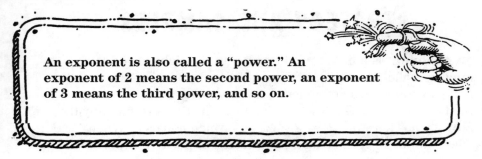

An exponent is also called a "power." An exponent of 2 means the second power, an exponent of 3 means the third power, and so on.

If you have two or more fractions, you can find powers two ways. For example, in $\left(\dfrac{1}{3}\right)^3 \cdot \left(\dfrac{3}{5}\right)^2$, you can use:

Marvelous Method 1

Multiply $\left(\frac{1}{3}\right)^3 \cdot \left(\frac{3}{5}\right)^2$.

1. Use the exponents to simplify each part of the problem separately.

$$\left(\frac{1}{3}\right)^3 = \frac{1}{3} \cdot \frac{1}{3} \cdot \frac{1}{3} = \frac{1}{27} \text{ and } \left(\frac{3}{5}\right)^2 = \frac{3}{5} \cdot \frac{3}{5} = \frac{9}{25}$$

2. Multiply $\frac{1}{27}$ and $\frac{9}{25}$.

3. Use canceling to divide both the 9 and 27 by 9, since 9 is a common factor.

$$\frac{1}{3\cancel{27}} \cdot \frac{\cancel{9}\,1}{25}$$

4. The problem now looks like $\frac{1}{3} \cdot \frac{1}{25}$, which is equal to $\frac{1}{75}$ by multiplying the numerators and denominators.

Or you can use:

Marvelous Method 2

Multiply $\left(\frac{1}{3}\right)^3 \cdot \left(\frac{3}{5}\right)^2$.

1. Use exponents to expand your problem, but don't simplify at this point. (You can work with smaller numbers this way.)

$$\left(\frac{1}{3}\right)^3 = \frac{1}{3} \cdot \frac{1}{3} \cdot \frac{1}{3} \text{ and } \left(\frac{3}{5}\right)^2 = \frac{3}{5} \cdot \frac{3}{5}$$

2. The problem looks like $\frac{1}{3} \cdot \frac{1}{3} \cdot \frac{1}{3} \cdot \frac{3}{5} \cdot \frac{3}{5}$.

3. While keeping the fractions in this form, use canceling to simplify your answer. Cancel the two 3s from the numerator with two of the 3s from the denominator.

$$\frac{1}{{}_1\cancel{3}} \cdot \frac{1}{{}_1\cancel{3}} \cdot \frac{1}{3} \cdot \frac{\cancel{3}^1}{5} \cdot \frac{\cancel{3}^1}{5}$$

4. The problem now looks like $\frac{1}{1} \cdot \frac{1}{1} \cdot \frac{1}{3} \cdot \frac{1}{5} \cdot \frac{1}{5}$, which is equal to $\frac{1}{75}$ by multiplying the numerators and denominators.

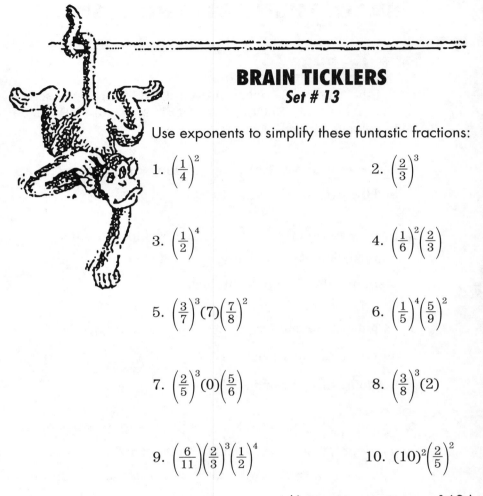

BRAIN TICKLERS
Set # 13

Use exponents to simplify these funtastic fractions:

1. $\left(\frac{1}{4}\right)^2$

2. $\left(\frac{2}{3}\right)^3$

3. $\left(\frac{1}{2}\right)^4$

4. $\left(\frac{1}{6}\right)^2\left(\frac{2}{3}\right)$

5. $\left(\frac{3}{7}\right)^3(7)\left(\frac{7}{8}\right)^2$

6. $\left(\frac{1}{5}\right)^4\left(\frac{5}{9}\right)^2$

7. $\left(\frac{2}{5}\right)^3(0)\left(\frac{5}{6}\right)$

8. $\left(\frac{3}{8}\right)^3(2)$

9. $\left(\frac{6}{11}\right)\left(\frac{2}{3}\right)^3\left(\frac{1}{2}\right)^4$

10. $(10)^2\left(\frac{2}{5}\right)^2$

(Answers are on page 162.)

BRAIN TICKLERS—THE ANSWERS

Set # 13, page 161

Most of the solutions here use MARVELOUS METHOD 2 (see page 160), but MARVELOUS METHOD 1 (see page 159) can be used as an alternative.

1. $\left(\frac{1}{4}\right)^2$ means to multiply $\frac{1}{4} \times \frac{1}{4}$. Multiply the two numerators and the two denominators to equal $\frac{1}{16}$.

2. $\left(\frac{2}{3}\right)^3$ means to multiply $\frac{2}{3} \times \frac{2}{3} \times \frac{2}{3}$. Multiply the numerators and the denominators, $\frac{2 \times 2 \times 2}{3 \times 3 \times 3}$, to equal $\frac{8}{27}$. A shortcut for doing this problem is $\left(\frac{2}{3}\right)^3 = \frac{2^3}{3^3} = \frac{8}{27}$.

3. $\left(\frac{1}{2}\right)^4$ means to multiply $\frac{1}{2} \times \frac{1}{2} \times \frac{1}{2} \times \frac{1}{2}$. Multiply the numerators and the denominators, $\frac{1 \times 1 \times 1 \times 1}{2 \times 2 \times 2 \times 2} = \frac{1}{16}$. A shortcut for doing this problem is $\left(\frac{1}{2}\right)^4 = \frac{1^4}{2^4} = \frac{1}{16}$.

4. Expand $\left(\frac{1}{6}\right)^2\left(\frac{2}{3}\right)$ to equal $\frac{1}{6} \cdot \frac{1}{6} \cdot \frac{2}{3}$. Since there is a common factor of 2 that divides evenly into the 2 in the numerator and one of the 6s in the denominator, you can use canceling to simplify, $\frac{1}{6} \cdot \frac{1}{{}_3 6} \cdot \frac{{}^1 2}{3}$. Multiply the remaining numerators and denominators to equal $\frac{1}{54}$.

5. Expand $\left(\frac{3}{7}\right)^3 (7)\left(\frac{7}{8}\right)^2$ to equal $\frac{3}{7} \cdot \frac{3}{7} \cdot \frac{3}{7} \cdot 7 \cdot \frac{7}{8} \cdot \frac{7}{8}$. Cancel the three 7s in the numerator with the three 7s in the denominator, $\frac{3}{{}_1 7} \cdot \frac{3}{{}_1 7} \cdot \frac{3}{{}_1 7} \cdot {}^1 7 \cdot \frac{7^1}{8} \cdot \frac{7^1}{8}$. Multiply the remaining numerators and denominators, $\frac{3 \times 3 \times 3}{8 \times 8}$, to equal $\frac{27}{64}$.

6. Expand $\left(\frac{1}{5}\right)^4\left(\frac{5}{9}\right)^2$ to $\frac{1}{5} \cdot \frac{1}{5} \cdot \frac{1}{5} \cdot \frac{1}{5} \cdot \frac{5}{9} \cdot \frac{5}{9}$. Cancel the two 5s in the numerator with two of the 5s in the denominator, $\frac{1}{1\cancel{5}} \cdot \frac{1}{1\cancel{5}} \cdot \frac{1}{5} \cdot \frac{1}{5} \cdot \frac{\cancel{5}^1}{9} \cdot \frac{\cancel{5}^1}{9}$. Multiply the remaining numerators and denominators, $\frac{1 \cdot 1 \cdot 1 \cdot 1}{5 \cdot 5 \cdot 9 \cdot 9}$, to equal $\frac{1}{2025}$.

7. When you multiply any whole number, fraction, or decimal by 0, your answer is 0.

8. Expand $\left(\frac{3}{8}\right)^3(2)$ to equal $\frac{3}{8} \times \frac{3}{8} \times \frac{3}{8} \times 2$. Cancel the 2 in the numerator with one of the 8s in the denominator, $\frac{3}{8} \times \frac{3}{8} \times \frac{3}{\cancel{8}_4} \times \cancel{2}^1$. Multiply your remaining numerators and denominators, $\frac{3 \times 3 \times 3}{8 \times 8 \times 4}$, to equal $\frac{27}{256}$.

9. Expand $\left(\frac{6}{11}\right)\left(\frac{2}{3}\right)^3\left(\frac{1}{2}\right)^4$ to $\left(\frac{6}{11}\right)\left(\frac{2}{3}\right)\left(\frac{2}{3}\right)\left(\frac{2}{3}\right)\left(\frac{1}{2}\right)\left(\frac{1}{2}\right)\left(\frac{1}{2}\right)\left(\frac{1}{2}\right)$. Cancel the three 2s in the numerator with three of the 2s in the denominator. Also, since 3 is a common factor to both 6 and 3, use canceling to divide 3 into both 6 and 3. Cancel the two remaining 2s. Notice that 6 is canceled to 2, then 2 is canceled to 1. $\frac{\cancel{\cancel{6}}^{1}}{11} \cdot \frac{\cancel{2}^1}{3} \cdot \frac{\cancel{2}^1}{3} \cdot \frac{\cancel{2}^1}{\cancel{3}_1} \cdot \frac{1}{\cancel{2}_1} \cdot \frac{1}{\cancel{2}_1} \cdot \frac{1}{\cancel{2}_1} \cdot \frac{1}{\cancel{2}_1}$. Now, multiply your remaining numerators and denominators, $\frac{1 \cdot 1 \cdot 1 \cdot 1}{11 \cdot 3 \cdot 3}$ to equal $\frac{1}{99}$.

If a number is used for canceling once, it can be used again in its new form as long as there is another common factor that divides into it.

10. Since 10 is a whole number, $10^2 = 10 \cdot 10 = 100$, and $\left(\frac{2}{5}\right)^2 = \frac{2}{5} \times \frac{2}{5} = \frac{4}{25}$. Now multiply $100 \cdot \frac{4}{25}$. Cancel the 25 and the 100, since 25 is a common factor to both 25 and 100, $\overset{4}{\cancel{100}} \cdot \frac{4}{\underset{1}{25}}$. Multiply $4 \cdot 4$ to get the answer of 16.

CHAPTER TEN

Outrageous Order
of Operations

If your numbers are in great disorder,
Using PEMDAS will put them in order.

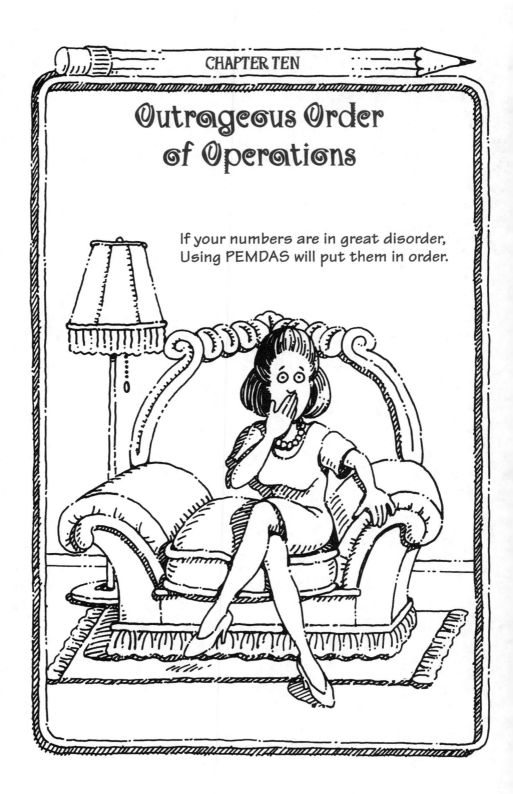

STRATEGY AND SAMPLES

Okay, here's the order of operations:

I'll have 2 additions and 1 multiplication to go, please.

No, **No**, the order of operations in mathematics refers to a prescribed way of doing a math problem. The order can be described as follows:

1. Perform all operations inside parentheses or with exponents as they occur from *left to right* in the problem.

2. Perform any multiplications or divisions as they occur from *left to right* in the problem.

3. Perform any additions or subtractions as they occur from *left to right* in the problem.

This order of operations can easily be remembered by the letters **PEMDAS** and the phrase "**P**lease **e**xcuse **m**y **d**ear **A**unt **S**ally."

Level 1	Level 2	Level 3
P E	M D	A S

These three levels form a hierarchy, as in a castle. For example, in a castle the highest level is the King or Queen, the middle level is the Prince or Princess, and the lowest level is non-royalty. First we must tend to the King and Queen. Next we take care of the Prince and Princess. Finally, we attend to the male and female non-royalty.

Both operations on each level are equal in importance to each other. Therefore, "as they occur from left to right" means that if division occurs before multiplication in a problem, you do the division first. (Who gets dinner first, the King or Queen? The rule is that whoever you *see* first gets his or her dinner first.) Likewise, do the same for each of the other levels.

For example, find $3 + 2 \times 4$.

1. In the order of operations, multiplication is performed before addition, so multiply the 2 times 4 first.

2. The problem now looks like $3 + 8$.

3. Perform the addition to find the answer of 11.

When fractions are involved, the same rules apply.

Orderly Combo #1

Find $\left(\frac{1}{2} + \frac{1}{3} \right) \times 2$.

1. Perform the operations inside the parentheses first. Therefore, add $\frac{1}{2} + \frac{1}{3}$ using the fraction rules. To do this, use the common denominator of 6 and convert each fraction to sixths. So, $\frac{1}{2} = \frac{3}{6}$ because $6 \div 2 = 3$ and $1 \cdot 3 = 3$, and $\frac{1}{3} = \frac{2}{6}$ because $6 \div 3 = 2$ and $1 \cdot 2 = 2$.

2. The problem now looks like $\frac{3}{6} + \frac{2}{6}$.

3. Add the numerators (in this case $3 + 2 = 5$).

4. The answer is $\frac{5}{6}$.

5. Replace the $\frac{1}{2} + \frac{1}{3}$ with $\frac{5}{6}$ in the original problem.

6. The problem now looks like $\frac{5}{6} \times 2$.

7. Use cancellation with multiplication to divide 2 into 6 and 2 into 2, $\frac{5}{{}_3 \cancel{6}} \times \cancel{2}^1$

8. The problem now looks like $\frac{5}{3} \times 1$.

9. Multiply. The answer is $\frac{5}{3}$ or $1\frac{2}{3}$.

Orderly Combo #2

Find $7\frac{7}{8} - \left(\frac{5}{6} \div \frac{1}{3}\right)^2$

1. By the order of operations, parentheses and exponents are on the same level (like the King and Queen), so since the parentheses occur first from left to right, perform the operations in the parentheses first.

2. Divide $\frac{5}{6}$ by $\frac{1}{3}$ by changing $\frac{1}{3}$ to its reciprocal and multiplying.

3. The part inside the parentheses looks like $\left(\frac{5}{6} \times \frac{3}{1}\right)$.

4. Use canceling to divide the 3 and the 6 by 3, $\frac{5}{{}_{2}\cancel{6}} \times \frac{\cancel{3}^{1}}{1}$.

5. You now have $\left(\frac{5}{2} \times \frac{1}{1}\right) = \frac{5}{2}$.

6. Square the $\frac{5}{2}$ to equal $\frac{25}{4}$.

$$\left(\frac{5}{2}\right)^2 = \frac{5}{2} \times \frac{5}{2} = \frac{25}{4}$$

7. Change the $\frac{25}{4}$ to the mixed number $6\frac{1}{4}$ by dividing 4 into 25.

8. Now substitute $6\frac{1}{4}$ back in the original problem in place of the parentheses and exponent.

9. The problem now looks like $7\frac{7}{8} - 6\frac{1}{4}$.

10. Use a common denominator of 8. Rewrite the $6\frac{1}{4}$ with a denominator of eighths, $6\frac{1}{4} = 6\frac{2}{8}$ because $8 \div 4 = 2$ and $1 \times 2 = 2$.

11. Perform the subtraction $7\frac{7}{8} - 6\frac{2}{8}$.

12. Keep the denominators and subtract the numerators, $\frac{7}{8} - \frac{2}{8} = \frac{5}{8}$.

13. Subtract the whole numbers ($7 - 6 = 1$). The answer is $1\frac{5}{8}$.

BRAIN TICKLERS
Set # 14

Use the order of operations to simplify these funtastic fractions:

1. $\left(\frac{5}{12} + \frac{1}{3}\right)^2$

2. $\left(3\frac{1}{2} - 2\right) \div 3 + 1\frac{3}{4}$

3. $1\frac{1}{3} + \frac{6}{7} \times \frac{14}{15}$

4. $16 - \left(\frac{1}{4} \cdot \frac{8}{11}\right) + \left(\frac{1}{2}\right)^2$

5. $\left(2\frac{1}{4} - \frac{1}{8}\right) \div \left(\frac{5}{8} + \frac{1}{4}\right)$

6. $\frac{1}{2} + 3\frac{1}{3} \times 6 \div \left(\frac{1}{3}\right)^2 - 1$

7. $\frac{1}{3} \div \frac{1}{2} + \frac{3}{4}$

8. $\left(\frac{3}{5}\right)^3 - \frac{3}{25}$

9. $\frac{3}{4}\left(\frac{4}{9}\right)^2 + \frac{1}{2}$

10. $\left(\frac{3}{4}\right)^2 \div \left(\frac{3}{8} - \frac{1}{12}\right)$

(Answers are on page 173.)

BRAIN TICKLERS—THE ANSWERS

Set # 14, page 172

Follow PEMDAS for all problems below.

1. Perform the addition operation *inside the parentheses*. Find a common denominator for 12 and 3. The LCM is 12. Rewrite $\frac{1}{3}$ with the new denominator of 12, $\frac{1}{3} = \frac{?}{12}$. Divide 3 into 12 and multiply the result by 1, $12 \div 3 = 4$ and $4 \times 1 = 4$. So, $\frac{1}{3} = \frac{4}{12}$. The fraction $\frac{5}{12}$ stays the same because it already has the common denominator of 12. Add $\frac{5}{12} + \frac{4}{12}$ to equal $\frac{9}{12}$. Reduce $\frac{9}{12}$ by dividing the common factor of 3 into both the numerator and the denominator, $\frac{\cancel{9}^{3}}{\cancel{12}_{4}}$. The fraction is $\frac{3}{4}$ in simplest form. The problem now becomes $\left(\frac{3}{4}\right)^2$. To *square* $\frac{3}{4}$, multiply $\frac{3}{4} \times \frac{3}{4}$. Your answer is $\frac{9}{16}$.

2. Perform subtraction *inside the parentheses* first. In $3\frac{1}{2} - 2$, subtract the whole numbers and bring along the fraction to your answer. So $3\frac{1}{2} - 2 = 1\frac{1}{2}$. The problem now looks like $1\frac{1}{2} \div 3 + 1\frac{3}{4}$. Using PEMDAS, *division* is the next operation to perform. This section of your problem is $1\frac{1}{2} \div 3$. Change $1\frac{1}{2}$ to the improper fraction of $\frac{3}{2}$. Now you have $\frac{3}{2} \div 3$. Change 3 to its reciprocal of $\frac{1}{3}$ and multiply, $\frac{3}{2} \cdot \frac{1}{3}$. Cancel the two 3s, $\frac{\cancel{3}^{1}}{2} \cdot \frac{1}{\cancel{3}_{1}}$. This part of your answer equals $\frac{1}{2}$. The original problem is now simplified to $\frac{1}{2} + 1\frac{3}{4}$. Now, your last operation to perform is *addition*. Find a common denominator for 2 and 4. The LCM is 4. Rewrite

$\frac{1}{2}$ with the common denominator of 4, $\frac{1}{2} = \frac{?}{4}$. Divide 2 into 4 and multiply the result by 1, $4 \div 2 = 2$ and $2 \cdot 1 = 2$. So, $\frac{1}{2} = \frac{2}{4}$. Now your problem looks like $\frac{2}{4} + 1\frac{3}{4}$. Add the two fractions and bring along the whole number, $\frac{2}{4} + 1\frac{3}{4} = 1\frac{5}{4}$. Change $\frac{5}{4}$ to the mixed number $1\frac{1}{4}$ by dividing 4 into 5 and getting a fractional remainder. You now have $1\frac{5}{4} = 1 + 1\frac{1}{4}$. Add your two whole numbers and bring along your fraction, $1 + 1\frac{1}{4} = 2\frac{1}{4}$. Your final answer is $2\frac{1}{4}$.

3. *Multiplication* is the first operation to perform here according to PEMDAS. In the section $\frac{6}{7} \times \frac{14}{15}$, you can divide 7 into both 7 and 14, since 7 is a common factor of both 7 and 14, $\frac{6}{_{1}7} \cdot \frac{14}{15}^{2}$. Since 3 is a common factor of both 6 and 15, divide 3 into both 6 and 15, $\frac{^{2}6}{1} \cdot \frac{2}{15_{5}}$. Multiply your numerators and denominators to get $\frac{4}{5}$. Now your problem looks like $1\frac{1}{3} + \frac{4}{5}$. *Add* by finding a common denominator for 5 and 3. The LCM is 15. Rewrite $\frac{4}{5}$ with the common denominator of 15, $\frac{4}{5} = \frac{?}{15}$. Divide 5 into 15 and multiply the result by 4, $15 \div 5 = 3$ and $3 \cdot 4 = 12$. Now, $\frac{4}{5} = \frac{12}{15}$. Similarly, $1\frac{1}{3} = 1\frac{5}{15}$. Now add $1\frac{5}{15} + \frac{12}{15}$. Bring along the whole number and add the two fractions, $1\frac{5}{15} + \frac{12}{15} = 1\frac{17}{15}$. Rewrite $\frac{17}{15}$ as the mixed number $1\frac{2}{15}$. Now you have $1 + 1\frac{2}{15}$. Add the two whole numbers and bring along the fraction, $1 + 1\frac{2}{15} = 2\frac{2}{15}$. The simplified answer is $2\frac{2}{15}$.

4. The first operation to perform here is the multiplication within the *parentheses*, $\left(\frac{1}{4} \cdot \frac{8}{11}\right)$. Since 4 is a common factor to both 4 and 8, you can use canceling to simplify, $\left(\frac{1}{_{1}4} \cdot \frac{8}{11}^{2}\right)$. Multiply what's left to get a result of $\frac{2}{11}$ in paren-

theses. The problem now looks like $16 - \frac{2}{11} + \left(\frac{1}{2}\right)^2$. The second operation to perform here involves the *exponent*, $\left(\frac{1}{2}\right)^2$. Multiply $\frac{1}{2} \cdot \frac{1}{2}$ to equal $\frac{1}{4}$. You now have $16 - \frac{2}{11} + \frac{1}{4}$. Now because addition and subtraction are on equal levels of importance, do what comes first in the problem from left to right. So, *subtraction* is next, $16 - \frac{2}{11}$. Change the 16 into $15\frac{11}{11}$ by subtracting 1 whole from the 16, making it a 15, and trading the 1 whole for $\frac{11}{11}$. Now subtract $15\frac{11}{11} - \frac{2}{11}$ by subtracting the two fractions and bringing along the whole number. The result of the subtraction is $15\frac{9}{11}$. The problem now looks like $15\frac{9}{11} + \frac{1}{4}$. Now, *addition* is the only operation left. Find a common denominator for 4 and 11. The LCM of 4 and 11 is 44. Rewrite $\frac{9}{11}$ as $\frac{?}{44}$ by dividing 11 into 44 and multiplying the result by 9, $44 \div 11 = 4$ and $4 \cdot 9 = 36$. So, $\frac{9}{11} = \frac{36}{44}$. Similarly $\frac{1}{4} = \frac{11}{44}$. The problem now looks like $15\frac{36}{44} + \frac{11}{44}$. Add the two numerators and bring along the whole number to get $15\frac{47}{44}$. Rewrite $\frac{47}{44}$ as the mixed number $1\frac{3}{44}$. You now have $15 + 1\frac{3}{44}$. Add to get $16\frac{3}{44}$.

5. First, work with subtraction inside the first *parentheses* from left to right, $(2\frac{1}{4} - \frac{1}{8})$. Find a common denominator for 4 and 8. The LCM is 8. Rewrite $2\frac{1}{4}$ as $2\frac{?}{8}$ by dividing 4 into 8 and multiplying the result by 1, $8 \div 4 = 2$ and $2 \cdot 1 = 2$. So, $2\frac{1}{4} = 2\frac{2}{8}$. Keep $\frac{1}{8}$ the same because it already has the common denominator of 8. Subtract $2\frac{2}{8} - \frac{1}{8}$ to equal $2\frac{1}{8}$. Now, work with addition inside the second *parentheses*. To add $\frac{5}{8} + \frac{1}{4}$ you need a common denominator of 8

as found above. Change $\frac{1}{4}$ into $\frac{2}{8}$ by the same procedure as above and add this to the $\frac{5}{8}$, $\frac{5}{8} + \frac{2}{8} = \frac{7}{8}$. The original problem is now changed to $2\frac{1}{8} \div \frac{7}{8}$. Your last operation is *division*. Rewrite $2\frac{1}{8}$ as the improper fraction $\frac{17}{8}$. Then change $\frac{7}{8}$ to its reciprocal of $\frac{8}{7}$ and multiply, $\frac{17}{8} \cdot \frac{8}{7}$. Now, the two 8s will cancel, $\frac{17}{1\cancel{8}} \cdot \frac{\cancel{8}^1}{7}$. Multiply your remaining numbers to get $\frac{17}{7}$. Change this improper fraction to a mixed number by dividing 7 into 17 with a fractional remainder. Your final simplified answer is $2\frac{3}{7}$.

6. According to PEMDAS, the *exponent* should be worked on first in this problem. Here, $\left(\frac{1}{3}\right)^2 = \frac{1}{3} \cdot \frac{1}{3}$. Multiply the numerators and denominators to equal $\frac{1}{9}$. The next operation in order is multiplication. Multiply $3\frac{1}{3} \times 6$. Rewrite $3\frac{1}{3}$ as the improper fraction $\frac{10}{3}$. You now have $\frac{10}{3} \cdot 6$. Since 3 is a common factor to both 3 and 6, you can use canceling to show the division, $\frac{10}{1\cancel{3}} \cdot \cancel{6}^2$. Multiply the remaining numbers to get 20. The original problem now is changed to $\frac{1}{2} + 20 \div \frac{1}{9} - 1$. The third operation here is *division*. The division part is $20 \div \frac{1}{9}$. Change $\frac{1}{9}$ to its reciprocal of 9 and change the problem to multiplication, $20 \cdot \frac{9}{1}$. Multiply to get 180. The original problem now looks like $\frac{1}{2} + 180 - 1$. Since addition and subtraction are on the same level, perform *addition* first because it occurs first from left to right. So, $\frac{1}{2} + 180 = 180\frac{1}{2}$, and then $180\frac{1}{2} - 1 = 179\frac{1}{2}$ by performing the final operation of *subtraction*.

7. Here *division* is the first operation to perform. In $\frac{1}{3} \div \frac{1}{2}$, invert $\frac{1}{2}$ to its reciprocal of $\frac{2}{1}$ and change the problem to multiplication, $\frac{1}{3} \times \frac{2}{1}$. Multiply to get $\frac{2}{3}$. Now you have the *addition* problem $\frac{2}{3} + \frac{3}{4}$. Find a common denominator for 3 and 4. The LCM is 12. Rewrite $\frac{2}{3}$ with the new denominator of 12, $\frac{2}{3} = \frac{?}{12}$, by dividing 3 into 12 and multiplying the result by 2 to get $\frac{2}{3} = \frac{8}{12}$. Similarly, $\frac{3}{4} = \frac{9}{12}$. Now add $\frac{8}{12} + \frac{9}{12}$. Your answer is $\frac{17}{12}$ by adding the two numerators. Rewrite $\frac{17}{12}$ as the mixed number $1\frac{5}{12}$. Your final answer is $1\frac{5}{12}$.

8. In this problem, simplify the portion with the *exponent* first. If you multiply your numerators and denominators, you get $\frac{27}{125}$. Your original problem now looks like $\frac{27}{125} - \frac{3}{25}$. Now do the *subtraction* by finding a common denominator for 25 and 125. The LCM is 125. Rewrite $\frac{3}{25}$ with the new denominator of 125, $\frac{3}{25} = \frac{?}{125}$, by dividing 25 into 125 and multiplying the result by 3. So, $\frac{3}{25} = \frac{15}{125}$. The fraction $\frac{27}{125}$ stays the same because it already has the common denominator. Set up your subtraction to look like $\frac{27}{125} - \frac{15}{125}$. Subtract the two numerators to get $\frac{12}{125}$.

9. The portion with the exponent $\left(\frac{4}{9}\right)^2$, can be simplified to $\frac{16}{81}$ by multiplying $4 \cdot 4$ for the numerator and $9 \cdot 9$ for the denominator. The problem now reads $\frac{3}{4} \cdot \frac{16}{81} + \frac{1}{2}$. Perform *multiplication* next, $\frac{3}{4} \cdot \frac{16}{81}$. Cancel the 3 from the numerator with the 81 in the denominator and, similarly, the 4 in the denominator with the 16 in the numerator,

$\frac{1}{1}\frac{\cancel{3}}{\cancel{4}} \cdot \frac{\cancel{16}^{4}}{\cancel{81}_{27}}$. The result is $\frac{4}{27}$. Now the problem looks like $\frac{4}{27} + \frac{1}{2}$. To perform the last operation of *addition*, find a common denominator for 27 and 2. The LCM is 54. Rewrite $\frac{4}{27}$ with the common denominator of 54 by dividing 27 into 54 and multiplying the result by 4, $\frac{4}{27} = \frac{8}{54}$. Similarly, $\frac{1}{2} = \frac{27}{54}$. Now add $\frac{8}{54} + \frac{27}{54}$. Add the two numerators to equal $\frac{35}{54}$.

10. Here, the operation with the *exponent*, $\left(\frac{3}{4}\right)^{2}$, is the first operation to perform, $\left(\frac{3}{4}\right)^{2} = \frac{3 \cdot 3}{4 \cdot 4} = \frac{9}{16}$. Now you have $\frac{9}{16} \div \left(\frac{3}{8} - \frac{1}{12}\right)$. Next work with the operation inside the *parentheses*. This is the subtraction problem $\frac{3}{8} - \frac{1}{12}$. Find a common denominator for 8 and 12. The LCM is 24. Rewrite $\frac{3}{8}$ into the new denominator of 24 by dividing 8 into 24 and multiplying this number by 3, $\frac{3}{8} = \frac{9}{24}$. Similarly, $\frac{1}{12} = \frac{2}{24}$. Subtract $\frac{9}{24} - \frac{2}{24}$. This part of the answer is $\frac{7}{24}$. Your original problem now looks like $\frac{9}{16} \div \frac{7}{24}$. The third and last operation to perform in this problem is *division*. Invert $\frac{7}{24}$ to its reciprocal of $\frac{24}{7}$. Set up the multiplication problem $\frac{9}{16} \cdot \frac{24}{7}$. The greatest common factor (GCF) of 16 and 24 is 8. Use canceling to divide 8 into both numbers, $\frac{9}{\cancel{16}_{2}} \cdot \frac{\cancel{24}^{3}}{7}$. Multiply the remaining numbers to get $\frac{27}{14}$. Rewrite $\frac{27}{14}$ as a mixed number by dividing 14 into 27. Your final answer is $1\frac{13}{14}$.

CHAPTER ELEVEN

Plenty of Practice

Here is more practice for you to do
Puzzles and problems and solutions, too.

EVEN MORE BRAIN TICKLERS

1. Reduce $\frac{20}{45}$ to simplest form.

   a) $\frac{5}{9}$ b) $\frac{4}{45}$ c) $\frac{4}{9}$ d) $\frac{10}{25}$

2. $4\frac{7}{8} + 7\frac{3}{4}$

   a) $11\frac{5}{8}$ b) $12\frac{5}{8}$ c) $11\frac{5}{16}$ d) $12\frac{3}{8}$

3. $3\frac{1}{2} \div 2$

   a) $1\frac{3}{4}$ b) $\frac{4}{7}$ c) $6\frac{1}{2}$ d) $1\frac{1}{2}$

4. $8 - 2\frac{3}{7}$

   a) $6\frac{3}{7}$ b) $6\frac{4}{7}$ c) $5\frac{5}{7}$ d) $5\frac{4}{7}$

5. $\left(\frac{3}{4}\right)^2 + \frac{1}{2} - \frac{1}{8}$

   a) $\frac{15}{16}$ b) $\frac{9}{8}$ c) $1\frac{1}{4}$ d) $2\frac{1}{2}$

6. Jim runs $1\frac{1}{2}$ miles on Monday, $2\frac{1}{3}$ miles on Tuesday, and $1\frac{5}{6}$ miles on Wednesday. How many miles did he run in all?

    a) $4\frac{5}{6}$ b) $5\frac{2}{3}$ c) $4\frac{1}{6}$ d) 5

7. Change $4\frac{5}{6}$ to an improper fraction.

    a) $\frac{17}{6}$ b) $\frac{180}{6}$ c) $\frac{36}{6}$ d) $\frac{29}{6}$

8. $5\frac{1}{3} \times 4\frac{1}{2}$

    a) $20\frac{1}{6}$ b) 22   c) 24   d) 23

9. $\left(\frac{1}{2}\right)^2 \times 4$

    a) $\frac{1}{4}$ b) 2   c) $\frac{1}{2}$ d) 1

10. $\frac{7}{8} \div 3$

    a) $\frac{7}{24}$ b) $\frac{21}{8}$ c) $\frac{29}{24}$ d) $\frac{24}{7}$

11. There are 584 attorneys at a convention.

    A) If $\frac{1}{4}$ of them are women, how many are women?

      a) 2336   b) 292   c) 144   d) 146

    B) How many are men?

      a) 438   b) 146   c) 242   d) 779

12. $\left(1\frac{1}{3} - \frac{2}{3}\right) \times \frac{3}{5}$

    a) $\frac{1}{5}$ b) 3   c) $\frac{2}{5}$ d) $\frac{2}{3}$

13. Rewrite $\frac{14}{3}$ as a mixed number.

    a) $4\frac{1}{3}$ b) $2\frac{3}{4}$ c) $3\frac{1}{4}$ d) $4\frac{2}{3}$

14. Which fraction is larger, $\frac{2}{3}$ or $\frac{8}{9}$?

    a) $\frac{8}{9}$ b) $\frac{2}{3}$

15. $\frac{3}{4} + 1\frac{7}{12}$

    a) $1\frac{11}{12}$ b) $2\frac{1}{3}$ c) $2\frac{7}{12}$ d) $2\frac{3}{8}$

16. If we have traveled $2\frac{1}{2}$ hours of a 6-hour trip, how many hours are left to travel?

    a) $5\frac{1}{2}$ b) $8\frac{1}{2}$ c) $4\frac{1}{2}$ d) $3\frac{1}{2}$

17. $\left(\frac{2}{3}\right)^3$

    a) $\frac{6}{9}$ b) $\frac{4}{9}$ c) $\frac{8}{27}$ d) $\frac{8}{3}$

18. Which fraction is larger, $\frac{1}{8}$ or $\frac{1}{9}$?

    a) $\frac{1}{9}$ b) $\frac{1}{8}$

19. 189 ounces of cereal will be separated into $4\frac{1}{2}$-ounce boxes equally. How many boxes of cereal will there be?

    a) 43   b) 42   c) 44   d) 50

20. Reduce $8\frac{16}{18}$ to simplest form.

    a) $8\frac{8}{9}$ b) $8\frac{2}{9}$ c) $8\frac{1}{8}$ d) $8\frac{6}{9}$

21. $15\frac{3}{5} - 11\frac{4}{5}$

    a) $3\frac{4}{5}$ b) $4\frac{1}{5}$ c) $4\frac{4}{5}$ d) $3\frac{1}{5}$

22. Maggie's gym class played volleyball for $\frac{1}{5}$ of the period, basketball for $\frac{1}{4}$ of the period, and jogged for $\frac{1}{6}$ of the period.

    A) What part of the gym period was spent on these activities?

      a) $\frac{13}{15}$ b) $\frac{37}{60}$ c) $\frac{25}{60}$ d) $\frac{39}{60}$

    B) What fractional part of the period is left over?

      a) $\frac{23}{60}$ b) $\frac{25}{60}$ c) $\frac{21}{60}$ d) $\frac{30}{60}$

23. There are 145 students going on a class trip.

A) If $\frac{2}{5}$ of the class are female students, how many are female?

    a) 50   b) 58   c) 56   d) 78

B) How many are male?

    a) 122   b) 58   c) 72   d) 87

24. $\frac{3}{4} = \frac{?}{28}$

    a) 6   b) 12   c) 14   d) 21

25. $\frac{3}{8} + 1\frac{1}{6} + 3\frac{5}{12}$

    a) $4\frac{23}{24}$ b) $4\frac{7}{12}$ c) $5\frac{1}{12}$ d) $4\frac{9}{24}$

26. $\frac{6}{11} - \frac{1}{3}$

    a) $\frac{7}{33}$ b) $\frac{5}{8}$ c) $\frac{5}{11}$ d) $\frac{17}{33}$

PLENTY OF PRACTICE

27. $\left(\frac{1}{2}\right)^2 - \left(\frac{1}{4} \div 2\right)$

 a) $\frac{1}{4}$ b) 2 c) $\frac{1}{8}$ d) $\frac{3}{8}$

28. Arrange in order from largest to smallest: $\frac{1}{6}, \frac{2}{9}, \frac{2}{3}$.

 a) $\frac{2}{3}, \frac{2}{9}, \frac{1}{6}$ b) $\frac{2}{3}, \frac{1}{6}, \frac{2}{9}$ c) $\frac{2}{9}, \frac{1}{6}, \frac{2}{3}$ d) $\frac{1}{6}, \frac{2}{9}, \frac{2}{3}$

29. $7\frac{6}{7} - 2\frac{3}{7}$

 a) $5\frac{7}{9}$ b) $9\frac{3}{7}$ c) $4\frac{3}{7}$ d) $5\frac{3}{7}$

30. Tim spends $\frac{1}{6}$ of his salary on entertainment and $\frac{1}{4}$ on clothes.

 A) What fraction of his salary does he spend on these two items?

   a) $\frac{5}{12}$ b) $\frac{1}{10}$ c) $\frac{1}{5}$ d) $\frac{1}{3}$

 B) What fractional part is left over for other things?

   a) $\frac{5}{12}$ b) $\frac{5}{6}$ c) $\frac{7}{12}$ d) $\frac{7}{10}$

31. Charla and Javier want to bake cupcakes for their youth group. They use 2 cups of flour for 24 cupcakes. At this same rate, how many cups of flour will be needed for 36 cupcakes?

 a) 5 cups b) 3 cups c) 2 cups d) $3\frac{1}{2}$ cups

32. If 6 pounds of bananas cost $1.86, how much will 1 pound cost?

   a) $.62   b) $.93   c) $.31   d) $.98

33. Solve this proportion: $\frac{5}{6} = \frac{12}{n}$.

   a) $14\frac{2}{5}$ b) 18   c) 10   d) $14\frac{1}{4}$

34. Simplify this ratio: $\frac{28 \; skiers}{42 \; skiers}$.

   a) $\frac{2}{3}$ b) $\frac{6}{7}$ c) $\frac{4}{7}$ d) $\frac{7}{6}$

35. 160 high school freshmen attended the football game last Saturday. 240 sophomores attended the game.

   A) What was the ratio of freshmen to sophomores?

   a) $\frac{1}{12}$ b) $\frac{2}{3}$ c) $\frac{7}{12}$ d) $\frac{2}{5}$

   B) What was the ratio of freshmen to the total number of freshmen and sophomores?

   a) $\frac{1}{2}$ b) $\frac{1}{4}$ c) $\frac{2}{3}$ d) $\frac{2}{5}$

(Answers are on page 195.)

PUZZLING FRACTIONS #1

Why is a comb like a fraction?

Add or subtract to solve each problem below. Then find your answer in the column on the right and notice the letter next to it. On the next page, write this letter above the line that corresponds to the number of the question. When completed, you will have the answer to the riddle. The first problem is done for you. All answers are in simplest form.

1. $\frac{5}{6} = \frac{5}{6}$

    $+\frac{1}{3} = \frac{2}{6}$

    $\frac{7}{6} = 1\frac{1}{6} = P$

2. $7\frac{1}{2}$

    $-2\frac{3}{4}$

3. $2\frac{3}{8}$

    $+\ \frac{3}{5}$

E	$3\frac{1}{6}$
T	$19\frac{1}{5}$
H	1
K	$10\frac{2}{7}$
A	$4\frac{3}{4}$
O	$\frac{2}{5}$
R	$16\frac{1}{5}$
B	$10\frac{1}{3}$
Y	$2\frac{39}{40}$
M	$13\frac{1}{3}$
P	$1\frac{1}{6}$

4.   6

    $+ 4\frac{1}{3}$
    _____

8. $\frac{8}{15}$

    $- \frac{2}{15}$
    _____

5.   8

    $- 4\frac{5}{6}$
    _____

9. $7\frac{2}{7}$

    $+ 3$
    _____

6. $7\frac{9}{10}$

    $+ 11\frac{3}{10}$
    _____

10. $16\frac{5}{6}$

    $- 3\frac{1}{2}$
    _____

7. $\frac{3}{7}$

    $+ \frac{4}{7}$
    _____

11. $12\frac{2}{5}$

    $+ 3\frac{4}{5}$
    _____

$\overline{}\ \overline{}\ \overline{}\ \overline{}\quad \overline{}\ \overline{}\ \overline{}\ \overline{}\quad \overline{}\ \overline{}\ \overline{}\ \overline{}\quad \overline{}\quad \overset{P}{\overline{}}\ \overline{}\ \overline{}\ \overline{}$

6 7 5 3   4 8 6 7   10 2 9 5   2   1 2 11 6

(Answers are on page 207.)

PUZZLING FRACTIONS #2

Why is a numerator like an elevator?

Multiply or divide to solve each problem below. Then find your answer in the column on the right and notice the letter next to it. On the next page, write this letter above the line that corresponds to the number of the question. When completed, you will have the answer to the riddle. The first one is done for you. All answers are in simplest form.

1. $\frac{3}{5} \times \frac{1}{2} = \frac{3}{10} = O$

2. $1\frac{1}{2} \times \frac{1}{4}$

3. $5 \div \frac{1}{3}$

4. $16\frac{1}{2} \div 2\frac{2}{3}$

E	$6\frac{3}{16}$
T	$15\frac{2}{3}$
G	96
O	$\frac{3}{10}$
H	$11\frac{3}{8}$
P	$1\frac{1}{9}$
Y	$\frac{3}{8}$
B	15

5. $7\frac{5}{6} \times 2$

6. $\frac{4}{9} \div \frac{2}{5}$

7. $10\frac{2}{3} \div \frac{1}{9}$

8. $4\frac{3}{8} \times 2\frac{3}{5}$

$$\overline{}\ \overline{}\ \overline{}\ \overline{} \quad \overline{}\ \overset{O}{\overline{}}\ \overline{}\ \overline{} \quad \overline{}\ \overset{O}{\overline{}}\ \overline{}\ \overset{O}{\overline{}} \quad \overline{}\ \overline{}\ \overline{} \quad \overline{}\ \overset{O}{\overline{}}\ \overline{}$$

584231587151584516

(Answers are on page 207).

PUZZLING FRACTIONS #3

What do a soldier and a math problem have in common?

Solve all problems below. Then find your answer in the column on the right and notice the letter next to it. Write this letter on each line at the bottom of page 194 that contains the number of that question. When completed, you will have the answer to the riddle. The first one is done for you. All answers are in simplest form.

1. $\frac{5}{9} + \frac{1}{2} = \frac{10}{18} + \frac{9}{18} = \frac{19}{18} = 1\frac{1}{18} = $ N

2. $3\frac{1}{4} \div \frac{1}{8}$

A $\quad 2\frac{1}{8}$

B $\quad 21\frac{37}{40}$

D $\quad \frac{5}{18}$

E $\quad \frac{1}{16}$

3. $10 - 3\frac{2}{5}$

4. $\left(\frac{3}{5}\right)^2 + \frac{1}{5}$

5. $2\frac{3}{7} \times \frac{7}{8}$

6. $18\frac{4}{5} + 3\frac{1}{8}$

7. $6 \times \frac{3}{4}$

8. $\frac{6}{18} - \frac{1}{18}$

9. $12 \div \frac{1}{5}$

10. $1\frac{1}{4} - \left(\frac{2}{3} - \frac{1}{3}\right)$

F	$\frac{11}{12}$
H	456
L	$1\frac{1}{2}$
N	$1\frac{1}{18}$
O	$\frac{14}{25}$
R	26
T	60
W	$4\frac{1}{2}$
Y	$6\frac{3}{5}$

11. At the end of a busy Saturday, Baker's Bakery had $\frac{1}{2}$ of a blueberry pie, $\frac{1}{4}$ of an apple pie, and $\frac{3}{4}$ of a pumpkin pie left. How much pie was left in total?

12. $\left(\frac{1}{4}\right)^3 \times 4$

13. There were 684 students at the Ridgewood High School football game. $\frac{2}{3}$ of the students were male. How many male students attended the game?

$$\overline{9}\ \overline{13}\ \overline{12}\ \overline{3}\quad \overline{6}\ \overline{4}\ \overline{9}\ \overline{13}\quad \overline{10}\ \overline{4}\ \overline{11}\ \overline{11}\ \overline{4}\ \overline{7}\quad \overset{N}{\overline{5}}\ \overline{1}\quad \overline{4}\ \overline{2}\ \overline{8}\ \overline{12}\ \overline{2}$$

(Answers are on page 207.)

ANSWERS TO EVEN MORE BRAIN TICKLERS

1. Look for a number that divides evenly into both the numerator and denominator. Find the GCF if possible. Here, both 20 and 45 are divisible by 5 (since they end in 5 or 0), $\frac{20 \div 5 = 4}{45 \div 5 = 9}$. Therefore, the reduced fraction is $\frac{4}{9}$. The answer is (c).

2. Find a common denominator for 8 and 4. The multiples of 8 are 8, 16, 24, 32 . . . , and the multiples of 4 are 4, 8, 12, 16, 20, 24 . . . . The LCM is 8. The fraction $4\frac{7}{8}$ stays the same because the denominator is already 8. Rewrite $7\frac{3}{4}$ with an equivalent fraction with a denominator of 8 by dividing 8 by 4 to get 2, and then multiplying this 2 by the 3 in the numerator to get 6. The equivalent mixed number is $7\frac{6}{8}$. The problem now looks like $4\frac{7}{8} + 7\frac{6}{8}$. Add the two whole numbers, $4 + 7 = 11$. Add the two fractions, $\frac{7}{8} + \frac{6}{8} = \frac{13}{8}$. You now have $11\frac{13}{8}$. Divide 13 by 8 to get $1\frac{5}{8}$. Add $1\frac{5}{8}$ to the whole number 11 to get $11 + 1\frac{5}{8} = 12\frac{5}{8}$. The answer is (b).

3. Change $3\frac{1}{2}$ into an improper fraction by multiplying the denominator by the whole number and adding the numerator, $2 \times 3 + 1 = 7$. Put the 7 over the 2 to form the fraction $\frac{7}{2}$. Now invert the 2 after the $\div$ symbol to its reciprocal of $\frac{1}{2}$, and change the division symbol into multiplication to look like $\frac{7}{2} \times \frac{1}{2}$. Multiply the numerators and denominators to get $\frac{7}{4}$. Divide 7 by 4 to get $1\frac{3}{4}$. The answer is (a).

4. In order to subtract $2\frac{3}{7}$ from 8, you must subtract 1 from 8 to equal 7. In this case, the 1 whole is traded for $\frac{7}{7}$. The problem now looks like $7\frac{7}{7} - 2\frac{3}{7}$. Subtract the two whole numbers, $7 - 2 = 5$ and the two fractions, $\frac{7}{7} - \frac{3}{7} = \frac{4}{7}$. Your answer is $5\frac{4}{7}$. The answer is (d).

5. This problem uses the order of operations PEMDAS. The exponent 2 tells you to square the $\frac{3}{4}$: $\frac{3}{4} \times \frac{3}{4} = \frac{9}{16}$. Now the problem looks like $\frac{9}{16} + \frac{1}{2} - \frac{1}{8}$. Find a common denominator for all three fractions at once. The multiples of 16 are 16, 32, 48 . . . , the multiples of 2 are 2, 4, 6, 8, 10, 12, 14, 16 . . . , and the multiples of 8 are 8, 16, 24 . . . . So, 16 is your LCM and your common denominator. The fraction $\frac{9}{16}$ stays the same since 16 is already your common denominator. Change $\frac{1}{2}$ to $\frac{8}{16}$ by dividing the two denominators, $16 \div 2 = 8$, and then multiplying by the numerator of 1, $8 \times 1 = 8$. Therefore, $\frac{1}{2} = \frac{8}{16}$. Change $\frac{1}{8}$ to $\frac{2}{16}$ by dividing the two denominators, $16 \div 8 = 2$, and then multiplying by the numerator of 1, $2 \times 1 = 2$, to get $\frac{1}{8} = \frac{2}{16}$. The problem now looks like $\frac{9}{16} + \frac{8}{16} - \frac{2}{16}$. Add the numerators $9 + 8 = 17$ and then subtract the numerator 2, $17 - 2 = 15$. Your answer is $\frac{15}{16}$. The answer is (a).

6. The phrase "in all" indicates addition. Add $1\frac{1}{2} + 2\frac{1}{3} + 1\frac{5}{6}$ by using a common denominator of 6. Rewrite $1\frac{1}{2}$ with the new common denominator of 6 by dividing the denominators, $6 \div 2 = 3$, and multiplying this 3 by the 1 in the numerator to get $1\frac{1}{2} = 1\frac{3}{6}$. Next, rewrite $2\frac{1}{3}$ with a denominator of 6 by dividing 6 by 3 to get 2 and then multiplying this 2 by the 1

in the numerator to get $2\frac{1}{3} = 2\frac{2}{6}$. Finally, $1\frac{5}{6}$ stays the same because it already has a denominator of 6. The problem now looks like $1\frac{3}{6} + 2\frac{2}{6} + 1\frac{5}{6}$. Add all the whole numbers to equal 4 and all of the fractions to equal $\frac{10}{6}$. You now have $4\frac{10}{6}$. Rewrite $\frac{10}{6}$ as the mixed number $1\frac{4}{6}$ by dividing 10 by 6 and then reducing the fraction to get $1\frac{2}{3}$. Add $4 + 1\frac{2}{3}$ to get $5\frac{2}{3}$. The answer is (b).

7. Multiply the denominator by the whole number and add the numerator, $6 \times 4 + 5 = 29$. Put this number over the original denominator to get a fraction of $\frac{29}{6}$. The answer is (d).

8. Change both $5\frac{1}{3}$ and $4\frac{1}{2}$ into improper fractions. Multiply 3 by 5 and then add 1 to get 16. Put 16 over 3 to get $\frac{16}{3}$. Multiply 2 by 4 and then add 1 to get $8 + 1 = 9$. Put 9 over 2 to get $\frac{9}{2}$. The problem now looks like $\frac{16}{3} \times \frac{9}{2}$. You can cancel in both diagonal directions. Both 16 and 2 are divisible by 2, $16 \div 2 = 8$ and $2 \div 2 = 1$. Both 3 and 9 are divisible by 3, $9 \div 3 = 3$ and $3 \div 3 = 1$. The problem now looks like $\frac{8}{1} \times \frac{3}{1}$. Multiply 8×3 to get 24. The answer is (c).

9. Expand $\left(\frac{1}{2}\right)^2$ to $\frac{1}{2} \times \frac{1}{2}$. Multiply the numerators and denominators to get $\frac{1}{4}$. Then multiply $\frac{1}{4} \times 4$. Cancel by dividing 4 into the denominator of 4 and into the whole number 4, $4 \div 4 = 1$. The problem now looks like $\frac{1}{1} \times 1$. The answer is 1 or (d).

10. Change the 3 to its reciprocal of $\frac{1}{3}$. Also change $\div$ to $\times$. The problem now looks like $\frac{7}{8} \times \frac{1}{3}$. Multiply the numerators and denominators to get $\frac{7}{24}$. The answer is (a).

11. A) The word "of" often indicates multiplication with fractions. Here, multiply $\frac{1}{4} \times 584$. Cancel by dividing 4 into 4 and 584. The problem now looks like $\frac{1}{1} \times 146$ or 146. The answer is (d)

    B) Since there are 584 attorneys altogether, subtract the number of women from 584 to equal the number of men. Here, $584 - 146 = 438$. There are 438 men. The answer is (a).

12. The order of operations tells you to perform the operations inside parentheses first. Since $1\frac{1}{3}$ and $\frac{2}{3}$ have the same denominator, you can subtract without changing denominators, but you must change the fractions so that the first fraction is larger than the second. There are two ways to do this. One way is to rewrite $1\frac{1}{3}$ as the improper fraction $\frac{4}{3}$. Then subtract, $\frac{4}{3} - \frac{2}{3} = \frac{2}{3}$. The other way is to represent 1 as $\frac{3}{3}$ and add this $\frac{3}{3}$ to the $\frac{1}{3}$ that's already there, $\frac{3}{3} + \frac{1}{3} = \frac{4}{3}$. Then subtract, $\frac{4}{3} - \frac{2}{3} = \frac{2}{3}$. The quantity inside the parentheses is $\frac{2}{3}$. Next, multiply $\frac{2}{3} \times \frac{3}{5}$. Use canceling to divide 3 into both 3s in the diagonal direction, $3 \div 3 = 1$. The problem now looks like $\frac{2}{1} \times \frac{1}{5}$. Multiply the numerators and denominators to get $\frac{2}{5}$. The answer is (c).

13. Divide 3 into 14. You get a whole number of 4 with a remainder of 2. This 2 becomes your numerator, and the divisor of 3 becomes your denominator to form the mixed number $4\frac{2}{3}$. The answer is (d).

14. To see which fraction is larger, cross multiply your two diagonals from bottom to top, $\overset{18}{}\frac{2}{3} \times \frac{8}{9}\overset{24}{}$ to get $9 \times 2 = 18$ and $3 \times 8 = 24$. Since 24 is larger than 18, the fraction $\frac{8}{9}$ is larger than the fraction $\frac{2}{3}$. The answer is (a).

15. Find a common denominator (LCM) for 4 and 12. The multiples of 4 are 4, 8, 12, 16 . . . , and the multiples of 12 are 12, 24, 36 . . . . The LCM, which is the common denominator for 4 and 12, is 12. Rewrite $\frac{3}{4}$ with a denominator of 12, $\frac{3}{4} = \frac{?}{12}$. Divide 4 into 12 and get 3 and then multiply this by the 3 in the numerator to get $\frac{3}{4} = \frac{9}{12}$. Because $1\frac{7}{12}$ already has the common denominator, it stays the same. The problem now looks like $\frac{9}{12} + 1\frac{7}{12}$. The whole number is 1. Add the two fractions, $\frac{9}{12} + \frac{7}{12} = \frac{16}{12}$. You now have $1\frac{16}{12}$. Change the $\frac{16}{12}$ into $1\frac{4}{12}$. You now have $1 + 1\frac{4}{12}$. Add the two whole numbers, $1 + 1 = 2$. You now have $2\frac{4}{12}$. Reduce $\frac{4}{12}$ to simplest form by dividing both 4 and 12 by 4, $4 \div 4 = 1$ and $12 \div 4 = 3$, to get a fraction of $\frac{1}{3}$. You now have $2\frac{1}{3}$. The answer is (b).

16. The phrase "are left" indicates subtraction. Your problem will look like $6 - 2\frac{1}{2}$. Subtract 1 from the 6 and change this into $\frac{2}{2}$ because 2 is the denominator in the original problem, and $\frac{2}{2}$ is equal to 1. Your problem now looks like $5\frac{2}{2} - 2\frac{1}{2}$. Subtract the two whole numbers and the two fractions, $5 - 2 = 3$ and $\frac{2}{2} - \frac{1}{2} = \frac{1}{2}$, to get an answer of $3\frac{1}{2}$. The answer is (d).

17. $\frac{2}{3}$ raised to the third power or "cubed" means to multiply $\frac{2}{3}$ by itself 3 times, $\frac{2}{3} \times \frac{2}{3} \times \frac{2}{3}$. Since there is no canceling, multiply the numerators and the denominators, $2 \times 2 \times 2 = 8$ and $3 \times 3 \times 3 = 27$. Your fraction is now $\frac{8}{27}$. Since this fraction is in simplest form (no number divides evenly into both the numerator and denominator), the answer is (c).

18. To see which fraction is larger, cross multiply your two diagonals from bottom to top, $\overset{9}{\frac{1}{8}} \times \overset{8}{\frac{1}{9}}$, $9 \times 1 = 9$ and $8 \times 1 = 8$. Since 9 is larger than 8, the fraction $\frac{1}{8}$ is larger than the fraction $\frac{1}{9}$. Also, when two fractions have 1 as their numerator, the one with the smaller denominator is the larger fraction. The answer is (b).

19. The phrase "separated into" gives the hint of division. Your problem will look like $189 \div 4\frac{1}{2}$. Change $4\frac{1}{2}$ to an improper fraction by multiplying 2 by 4 and adding 1 to get 9. Put the 9 over the original denominator to get $\frac{9}{2}$. Invert $\frac{9}{2}$ to $\frac{2}{9}$. Then change your problem to multiplication. It now looks like $189 \times \frac{2}{9}$. Cancel by dividing 9 into 189 and into 9, $189 \div 9 = 21$ and $9 \div 9 = 1$. The problem now looks like $21 \times \frac{2}{1}$. Multiply 21×2 to get 42. The answer is (b).

20. To reduce $\frac{16}{18}$, look for a common factor that divides evenly into both 16 and 18. It is best to find the greatest common factor (GCF). The factors of 16 are 2, 4, 6, 8, 16, and the factors of 18 are 2, 3, 6, 9, 18. The GCF is 2. Divide both 16 and 18 by 2, $16 \div 2 = 8$ and $18 \div 2 = 9$. The reduced fraction is $\frac{8}{9}$. Bring along the whole number 8 to your answer. You now have $8\frac{8}{9}$. The answer is (a).

21. Since the denominators are the same, you do not have to find a common denominator. However, since the second fraction is larger than the first, you must change the first fraction so that its numerator is larger than the second. Subtract 1 from 15 to equal 14. In this case, the 1 is traded for $\frac{5}{5}$, since 5 is the denominator in the problem. Add this $\frac{5}{5}$ to the $\frac{3}{5}$ that is already there, $\frac{5}{5} + \frac{3}{5} = \frac{8}{5}$. The problem now looks like $14\frac{8}{5} - 11\frac{4}{5}$. Subtract the two whole numbers, $14 - 11 = 3$. Subtract the two fractions, $\frac{8}{5} - \frac{4}{5} = \frac{4}{5}$. The answer is $3\frac{4}{5}$, or (a).

22. A) Since three activities with three fractions are mentioned, the problem lends itself to addition. The fractions should be "totaled" even though the word "total" isn't mentioned. The fractions of $\frac{1}{5} + \frac{1}{4} + \frac{1}{6}$ need a common denominator in order to be added. The multiples of 5 are 5, 10, 15, 20, 25, 30, 35, 40, 45, 50, 55, 60 . . . , the multiples of 4 are 4, 8, 12, 16, 20, 24, 28, 32, 36, 40, 44, 48, 52, 56, 60 . . . , and the multiples of 6 are 6, 12, 18, 24, 30, 36, 42, 48, 54, 60 . . . . The LCM is 60. Rewrite $\frac{1}{5}$ as $\frac{?}{60}$ by dividing 60 by 5 to get 12 and then multiplying 12 by the 1 in the numerator to get 12. The equivalent fraction is $\frac{12}{60}$. Rewrite $\frac{1}{4}$ as $\frac{?}{60}$ by dividing 60 by 4 to get 15 and then multiplying 15 by the 1 in the numerator to get 15. The equivalent fraction is $\frac{15}{60}$. Rewrite $\frac{1}{6}$ as $\frac{?}{60}$ by dividing 60 by 6 to get 10 and then multiplying 10 by the 1 in the numerator to get 10. The equivalent fraction is $\frac{10}{60}$. The problem now looks like $\frac{12}{60} + \frac{15}{60} + \frac{10}{60}$. Add the numerators to get $\frac{37}{60}$. The answer is (b).

B) The phrase "left over" implies subtraction. When you talk about fractional parts, all of your parts must add up to 1. To find the part left over, you find the difference between 1 and $\frac{37}{60}$. Here, one whole is equal to $\frac{60}{60}$. Subtract $\frac{60}{60} - \frac{37}{60}$ to get $\frac{23}{60}$. The answer is (a).

23. A) Looking at the phrase "$\frac{2}{5}$ of the class," you are asked to find a fractional part of a given amount, which indicates multiplication. Here the part is $\frac{2}{5}$ females and the given amount is 145, so you multiply $\frac{2}{5} \cdot 145$. Use canceling to divide 5 into 145 and into itself, $5 \div 5 = 1$ and $145 \div 5 = 29$. The problem now looks like $\frac{2}{1} \cdot 29$. Multiply $2 \cdot 29$ to equal 58. So, 58 students are female. The answer is (b).

B) The easiest way to find the number of males is to subtract the number of females from the total number of students, $145 - 58 = 87$. There are 87 males. The answer is (d).

24. Divide the denominator 4 into the denominator 28 to get 7. Multiply this 7 by the numerator 3 to get 21. Your new equivalent fraction is $\frac{21}{28}$, and the answer is 21, or (d).

25. Find a common denominator that 8, 6, and 12 all divide into evenly. The multiples of 8 are 8, 16, 24 . . . , the multiples of 6 are 6, 12, 18, 24 . . . , and the multiples of 12 are 12, 24 . . . . The common denominator, which is the least common multiple, is 24. Now, $\frac{3}{8} = \frac{9}{24}$ by dividing 8 into 24 to get 3 and multiplying this 3 by the numerator of 3 to get 9; $1\frac{1}{6} = 1\frac{4}{24}$ by dividing 6 into 24 to get 4 and multiplying this 4 by the numerator of 1 to get 4; and $3\frac{5}{12} = 3\frac{10}{24}$ by dividing

12 into 24 to get 2 and multiplying this 2 by the numerator of 5 to get 10. Your problem now looks like $\frac{9}{24} + 1\frac{4}{24} + 3\frac{10}{24}$. Add the whole numbers, $1 + 3 = 4$, and the numerators of your fractions, $\frac{9}{24} + \frac{4}{24} + \frac{10}{24} = \frac{23}{24}$. The answer is $4\frac{23}{24}$, or (a).

26. Find a common denominator that 11 and 3 divide into evenly. The multiples of 11 are 11, 22, 33, 44 . . . , and the multiples of 3 are 3, 6, 9, 12, 15, 18, 21, 24, 27, 30, 33 . . . . The least common multiple, the common denominator, is 33. Now, $\frac{6}{11} = \frac{18}{33}$ by dividing 11 into 33 to get 3 and multiplying this 3 by the numerator of 6 to get 18; and $\frac{1}{3} = \frac{11}{33}$ by dividing 3 into 33 to get 11 and multiplying this 11 by the numerator of 1 to get 11. The problem now looks like $\frac{18}{33} - \frac{11}{33}$. Subtract the numerators of your fractions to get $\frac{7}{33}$. The answer is (a).

27. According to the order of operations, $\left(\frac{1}{2}\right)^2$ would be simplified first as $\frac{1}{2} \times \frac{1}{2} = \frac{1}{4}$. Next, work in the parentheses and invert 2 to its reciprocal of $\frac{1}{2}$ and multiply, $\frac{1}{4} \times \frac{1}{2}$. This result in the parentheses is $\frac{1}{8}$. The problem now looks like $\frac{1}{4} - \frac{1}{8}$. Find a common denominator for 4 and 8. The multiples of 4 are 4, 8, 12 . . . , and the multiples of 8 are 8, 16, 24 . . . . The LCM is 8. Use this for your common denominator. Now $\frac{1}{4} = \frac{2}{8}$ by dividing 4 into 8 to get 2 and multiplying this 2 by the 1 in the old numerator; and $\frac{1}{8}$ stays the same. The problem now looks like $\frac{2}{8} - \frac{1}{8}$. Subtract the numerators to get $\frac{1}{8}$ or (c).

28. To compare the size of fractions you must find a common denominator for all of the given fractions. The multiples of 6 are 6, 12, 18, 24 . . . , the multiples of 9 are 9, 18, 27 . . . , and the multiples of 3 are 3, 6, 9, 12, 15, 18 . . . . 18 is the LCM. Now, $\frac{1}{6} = \frac{3}{18}$ by dividing 6 into 18 to get 3 and multiplying this 3 by the numerator of 1 to get 3 as the new numerator; and $\frac{2}{9} = \frac{4}{18}$ by dividing 9 into 18 to get 2 and multiplying this 2 by the 2 in the old numerator; and $\frac{2}{3} = \frac{12}{18}$ by dividing 3 into 18 to get 6 and multiplying this 6 by the 2 in the old numerator to get 12 as the new numerator. Since $\frac{1}{6} = \frac{3}{18}, \frac{2}{9} = \frac{4}{18}$, and $\frac{2}{3} = \frac{12}{18}$, the largest fraction is $\frac{12}{18}$, or $\frac{2}{3}$. The next largest fraction is $\frac{4}{18}$, or $\frac{2}{9}$. The smallest fraction is $\frac{3}{18}$, or $\frac{1}{6}$. The fractions from largest to smallest are $\frac{2}{3}, \frac{2}{9}, \frac{1}{6}$, which is answer (a).

29. Since these two fractions already have a common denominator of 7, you can subtract your two fractions, $\frac{6}{7} - \frac{3}{7} = \frac{3}{7}$, and your two whole numbers, $7 - 2 = 5$. The answer is $5\frac{3}{7}$, or (d).

30. A) Since you do not know a salary amount, you can only work with the two fractions $\frac{1}{6}$ and $\frac{1}{4}$. Since the question mentions "on these two items," add $\frac{1}{6}$ and $\frac{1}{4}$ to find the fractional salary part spent. Find a common denominator. The multiples of 6 are 6, 12, 18 . . . , and the multiples of 4 are 4, 8, 12 . . . . The least common multiple, to use for the common denominator, is 12. Now $\frac{1}{6} = \frac{2}{12}$ by dividing 6 into 12 to get 2 and multiplying this 2 by the old numerator of 1 to get a new numerator of 2; and $\frac{1}{4} = \frac{3}{12}$ by dividing 4 into 12

to get 3 and multiplying this 3 by the old denominator of 1 to get a new numerator of 3. You now have $\frac{2}{12} + \frac{3}{12}$. Add to get $\frac{5}{12}$, or (a).

B) The words "left over" imply subtraction. When you talk about fractional parts, all of your parts must add up to 1 whole. To find the part left over, find the difference between 1 whole and $\frac{5}{12}$. Here, one whole is equal to $\frac{12}{12}$. Subtract $\frac{12}{12} - \frac{5}{12}$ to get $\frac{7}{12}$. The answer is (c).

31. Set up your problem as a proportion as follows: $\frac{cups\ of\ flour}{cupcakes} = \frac{cups\ of\ flour}{cupcakes}$. Replace the number of cups of flour and cupcakes from the problem into the appropriate places in your proportion: $\frac{2}{24} = \frac{c}{36}$ (2 cups of flour and 24 cupcakes go together and the unknown number of cups of flour c goes with 36 cupcakes). Multiply the cross products to get $24 \cdot c = 2 \cdot 36$, or $24c = 72$. Divide 24 into 72 to get an answer of 3. Therefore, at this rate, Charla and Javier would need 3 cups of flour for 36 cupcakes. This is equivalent to 2 cups of flour for 24 cupcakes. The correct answer is (b).

32. Since 1 pound indicates a unit rate, divide the number of pounds into the total price given to determine the unit price, or the price per pound $6\overline{)1.86}^{\,0.31}$. So, \$.31 is the price for 1 pound of bananas. The correct answer is (c).

33. Multiply $5 \cdot n$ and set this equal to $6 \cdot 12$, $5n = 72$. Divide 5 into 72 to get an answer of $14\frac{2}{5}$. Therefore $n = 14\frac{2}{5}$. The correct answer is (a).

34. Look for a common factor, preferably the GCF, that will divide evenly into both 28 and 42. 7 is a common factor that will divide evenly into both 28 and 42. Use canceling to simplify, $\frac{28}{42}\,^{4}_{6}$. Now, you can use a common factor of 2 to simplify $\frac{4}{6}$ to $\frac{2}{3}$. From the beginning, if you had noticed that the GCF for 28 and 42 is 14, you could have simplified the fraction in one step $\frac{28}{42}\,^{2}_{3}$, by dividing 14 into both the numerator and denominator. The correct answer is (a).

35. A) Set up the ratio as $\frac{number\ of\ freshmen}{number\ of\ sophomores}$. The number of freshmen is equal to 160 and the number of sophomores is 240, so your fractional ratio becomes $\frac{160}{240}$. Since 80 is the GCF, divide both 160 and 240 by 80, $\frac{160}{240}\,^{2}_{3}$. Therefore the ratio of freshmen to sophomores is $\frac{2}{3}$. The correct answer is (b).

B) Set up the ratio as $\frac{number\ of\ freshmen}{total\ number\ of\ freshmen\ and\ sophomores}$. The total number of freshmen and sophomores is equal to $160 + 240 = 400$. The number of freshmen is 160, so $\frac{160}{400}$ is your fractional ratio. Since 160 and 400 are both divisible by 80, simplify to $\frac{160}{400}\frac{2}{5}$. Therefore, the ratio of the number of freshmen to the total number of freshmen and sophomores is $\frac{2}{5}$. The correct answer is (d).

ANSWERS TO PUZZLING FRACTIONS

Puzzling Fractions #1

They both make a part!

Puzzling Fractions #2

They both go to the top!

Puzzling Fractions #3

They both follow an order!

INDEX